Representing and Reasoning with Qualitative Preferences

Tools and Applications

Synthesis Lectures on Artificial Intelligence and Machine Learning

Editors
Ronald J. Brachman, *Yahoo! Labs*
William W. Cohen, *Carnegie Mellon University*
Peter Stone, *University of Texas at Austin*

Representing and Reasoning with Qualitative Preferences: Tools and Applications
Ganesh Ram Santhanam, Samik Basu, and Vasant Honavar

ISBN: 978-3-031-00445-2   paperback
ISBN: 978-3-031-01573-1   ebook

DOI 10.1007/978-3-031-01573-1

A Publication in the Springer series
SYNTHESIS LECTURES ON ARTIFICIAL INTELLIGENCE AND MACHINE LEARNING

Lecture #31
Series Editors: Ronald J. Brachman, *Yahoo! Labs*
                William W. Cohen, *Carnegie Mellon University*
                Peter Stone, *University of Texas at Austin*
Series ISSN
Print 1939-4608   Electronic 1939-4616

Representing and Reasoning with Qualitative Preferences

Tools and Applications

Ganesh Ram Santhanam
Iowa State University

Samik Basu
Iowa State University

Vasant Honavar
Pennsylvania State University

SYNTHESIS LECTURES ON ARTIFICIAL INTELLIGENCE AND MACHINE LEARNING #31

ABSTRACT

This book provides a tutorial introduction to modern techniques for representing and reasoning about qualitative preferences with respect to a set of alternatives. The syntax and semantics of several languages for representing preference languages, including CP-nets, TCP-nets, CI-nets, and CP-theories, are reviewed. Some key problems in reasoning about preferences are introduced, including determining whether one alternative is preferred to another, or whether they are equivalent, with respect to a given set of preferences. These tasks can be reduced to model checking in temporal logic. Specifically, an induced preference graph that represents a given set of preferences can be efficiently encoded using a Kripke Structure for Computational Tree Logic (CTL). One can translate preference queries with respect to a set of preferences into an equivalent set of formulae in CTL, such that the CTL formula is satisfied whenever the preference query holds. This allows us to use a model checker to reason about preferences, i.e., answer preference queries, and to obtain a justification as to why a preference query is satisfied (or not) with respect to a set of preferences. This book defines the notions of the equivalence of two sets of preferences, including what it means for one set of preferences to subsume another, and shows how to answer preferential equivalence and subsumption queries using model checking. Furthermore, this book demontrates how to generate alternatives ordered by preference, along with providing ways to deal with inconsistent preference specifications. A description of CRISNER—an open source software implementation of the model checking approach to qualitative preference reasoning in CP-nets, TCP-nets, and CP-theories is included, as well as examples illustrating its use.

KEYWORDS

preferences, qualitative preferences, preference reasoning, model checking, knowledge representation, automated inference, decision support systems

We dedicate this book to our respective parents, spouses, children, teachers, mentors, and students.

Ganesh Ram Santhanam, Samik Basu, and Vasant Honavar

Contents

Acknowledgments

The research on which this book is based was conducted at the department of Computer Science at Iowa State University and the College of Information Sciences and Technology at the Pennsylvania State University and was supported in part by the grants CCF 1143734 and CCF 0702758 from the National Science Foundation and by the Edward Frymoyer Endowed Chair held by Vasant Honavar.

The authors are thankful to Zachary Oster at the Department of Computer Science, University of Wisconsin-Whitewater for his contribution to the tool during his doctoral research at Iowa State University and for his help with work implementing algorithms presented in Chapter 6. The authors are grateful to Ronen Brafman at the Department of Computer Science, Ben-Gurion University, Israel for helping us clarify the semantics of CP-net and TCP-net formalisms. The authors are grateful to the anonymous reviewers whose comments helped improve the manuscript and Mr. Michael B. Morgan of Morgan & Claypool Publishers and Dr. C.L. Tondo for their assistance to us with the manuscript preparation and publication process.

Ganesh Ram Santhanam, Samik Basu, and Vasant Honavar
December 2015

CHAPTER 1

Qualitative Preferences

The ability to represent and reason about preferences over a set of alternatives is central to rational decision making. For example, suppose Jane wants a low-rent apartment in midtown Manhattan, but no such apartment is available. Jane then has to settle for an apartment that is the most preferred with respect to her personal preferences from among the ones that are available. To complicate matters, Jane might have preferences with respect to the neighborhood, cost, and many other attributes of the apartment (e.g., the number of bedrooms, availability of parking, whether the apartment is on the ground floor, etc.). Furthermore, preferences with respect to some of the attributes, e.g., cost, may be more important than preferences with respect to other attributes.

Preferences have been the subject of study in many disciplines including economics, decision theory, social choice, and game theory. Artificial intelligence (AI) brings a fresh perspective to the study of preferences. Specifically, AI is concerned with how to efficiently represent and reason about preferences especially when the preferences involve multiple attributes, resulting in a large combinatorial space of alternatives.

Preferences over a set of alternatives can be qualitative (e.g., relative ordering among alternatives) or quantitative (e.g., numeric evaluation of each alternative) in nature. Qualitative preferences are more general in the sense that they define a *preference relation* (typically some kind of a partial order) over the set of alternatives, while quantitative preferences require more precision and map the set of alternatives to a numeric scale through a utility function [38]. Techniques for manipulating quantitative preferences have been well studied in the literature [54]. The seminal work of von Neumann and Morgenstern [62] provided an axiomatization of constraints over *rational* preferences and proved the existence of a *quantitative utility function* that maps each outcome to a value on a numeric scale as a consequence of the resulting axioms. In settings where quantitative preferences are available, the utility theory framework defines a utility function [38, 54] that can be used to decide the most preferred outcomes based on principles such as the *maximization of expected utility*.

However, in most real-world settings, quantitative preferences are simply unavailable. It is often easier to elicit qualitative preferences [36, 91]: the agent simply needs to be able to rank the alternatives. This is especially true in the case of conditional preferences. For example, while Jane prefers a job with a high salary if she lives in Manhattan and can't work from home, she is willing to settle for a lower salary if she lives on Long Island and can work from home. It is problematic to map qualitative preferences onto a numeric scale, because doing so is tantamount to injecting information that is simply unavailable in the preferences expressed by the agent, and hence distorts the agent's preferences. Hence, much of the recent work on preference representation and reasoning in AI focuses on qualitative preferences [35, 36, 40].

This book aims to provide a tutorial introduction to modern techniques for reasoning about qualitative preferences with respect to a set of alternatives. Specifically, the focus will be on reasoning about the qualitative preferences of a single agent[1] based on model checking, a technique that has been developed and successfully used at industry-scale for software program verification.

1   MOTIVATING EXAMPLES

We proceed to describe some real-world examples of scenarios that call for techniques for representing and reasoning about qualitative preferences.

1.1   CYBERDEFENSE POLICY

Network security management involves balancing between making the network maximally available and safeguarding against attacks that exploit the vulnerabilities exposed due to the services provided. The objective of a systems administrator or a cyberdefender managing a network is to take preventive measures that mitigate the risk of vulnerabilities in the system being exploited maliciously. Each vulnerability is typically associated with a set of attributes[2] that describe the potential risks and threats that it poses to the network. The job of a cyberdefender is then to make an informed decision as to which policies address vulnerabilities that pose the most serious threats. In order to design a good defense policy, the cyberdefender would have to prioritize the vulnerabilities based on the *seriousness* of threats they pose in terms of these attributes.

Consider three attributes describing the seriousness of threats posed by a vulnerability, namely (a) attack complexity (A) with values "High" or "Low" (indicating whether the complexity of attack required to exploit the vulnerability is high or low); (b) availability of exploit (E) with values "Code-available" or "Code-unavailable" (indicating whether easy-to-use exploit code is freely available or entirely theoretical); and (c) availability of fix (F) for the vulnerability with values "Fix-Available" or "Fix-Unavailable" (indicating whether a fix is readily available or not).

The defender may consider certain vulnerabilities to be more serious than others, based on certain priorities and tradeoffs over these attributes. For instance,

p_1 A vulnerability whose exploit code is known to be available ($E = $ Code-Available) is of higher priority than one whose exploitability has not been proven yet ($E = $ Code-Unavailable).

---

[1]Multi-agent decision making scenarios present additional challenges: Agents might have conflicting preferences. For example, while Jane wants a low-rent apartment in midtown Manhattan, Jane's husband John prefers a single family home on Long Island. Further complications arise from uncertainty associated with the preferences or choices. However, preference reasoning in the multi-agent setting and the treatment of uncertainty are beyond the scope of this book.

[2]The attributes may be provided by a database that maintains a list of known vulnerabilities and their characteristics such as the National Vulnerability Database [66].

p_2 Vulnerabilities with low access complexity (A = Low) are "preferred" (more serious) compared to those with High Access Complexity (A = High).

p_3 Among vulnerabilities whose exploit codes are available (E = Code-Available), those for which fixes are unavailable (F = Fix-Unavailable) are more serious than those for which official fixes are available (F = Fix-Available).

p_2' Attack Complexity is more important in determining the seriousness of a vulnerability; and it can be traded against the availability of exploit (E) in deciding which vulnerability is more serious.

Note that, if as per the preferences[3] described above, one vulnerability is preferred to another, then the former is a more serious vulnerability possibly requiring the attention of the cyberdefender. The above preferences can be represented using compact graphical representation languages such as CP-nets and TCP-nets; a CP-net for the preferences $p_1 \ldots p_3$ is shown in Figure 1.1(a), and a TCP-net for all the above preferences is shown in Figure 1.1(b). The objective of the defender is to prioritize the vulnerabilities that exist in the network based on the above preferences. The priorities induced by the preferences p_1 to p_3 are depicted as a directed graph in Figure 1.1(c) where the edges point toward vulnerabilities with higher priority (how these graphs are computed will be described in detail in Chapter 4). The defense policy is to "handle" a vulnerability that is higher in the priority before the vulnerability that is lower in the priority ordering; and hence obtaining an ordering of vulnerabilities based on their attributes is critical for providing optimal cyberdefense.

1.2 EDUCATION

Consider a student who has just enrolled in a Masters degree in Sustainability at the Department of Environmental Engineering in a university. The student is required to develop a program of study (POS) that is a set of courses chosen from the course catalog consisting of all offered graduate courses in the department. Each course may be classified into a subfield within Sustainability, such as Climate, Wildlife, Environmental Economics, or Energy. The catalog also lists the instructor and the number of credit hours each course is worth. The question of interest here is: Among all sets of courses that satisfy the requirement of a POS, which is the most preferred for the student?

Computing the preferred POS requires accounting for a student's preferences over the various courses, expressed in terms of preferences over the attributes of the courses. For instance, the student may prefer one instructor over another based on the student's past experience or ratings of the instructors he has gathered; or he may prefer courses with lower credit hours (higher number of credit hours may indicate more difficulty); or the student may prefer to specialize in a particular

---

[3]The usage of the numbering p_2' and p_2'' will be clear in the subsequent chapters. Intuitively, both p_2' and p_2'' further qualify the preference p_2 as they discuss the condition under which p_2 should be the deciding factor irrespective of the values of other attributes such as E in p_2', and E and F in p_2''.

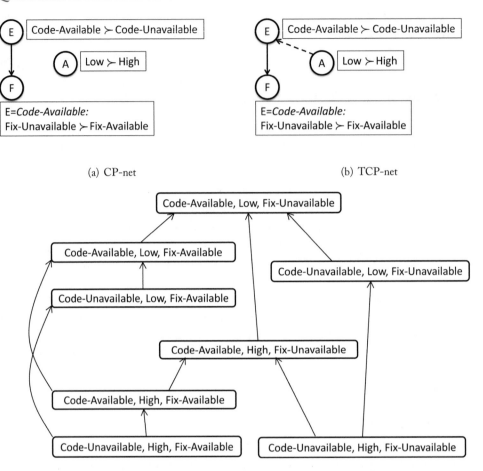

(a) CP-net   (b) TCP-net

(c) Priorities induced on vulnerabilities by the preferences p_1 to p_3 represented as a CP-net in Figure 1.1(a)

Figure 1.1: Risk priorities of a cyberdefender.[a]

[a]Formal syntax and semantics for CP-nets and TCP-nets will be given in Chapter 2.

subfield over other subfields of Sustainability, and hence, for example, prefer courses in Climate or Wildlife to Environmental Economics or Energy related courses. Given these preferences over the course attributes, the student may want to select the most preferred courses from the catalog to fill in his program of study, which involves reasoning with qualitative preferences. Another question is that given two programs of study consisting of different sets of courses, the student may want to know whether he should pick one in favor of the other. This brings up another interesting aspect of reasoning with qualitative preferences—how would the student reason about preferences over collections of objects (e.g., courses in this example)?

1.3   SOFTWARE ENGINEERING

Consider the design of a software application that allows customers to search for and order books, pay for the books that they ordered, and get the books shipped to a specified address. Functional requirements specify the desired functionality of software; for instance, searching for books, adding them to a cart, checking out using alternative payment options and choosing a shipping method. There are multiple implementations that could realize the desired functionality. The question of interest to us from the point of view of preference reasoning is: Which is the most preferred realization of the software design?

Answering the above question entails representing and reasoning with the preferences of users (software architects), who might have preferences over alternative implementations based on non-functional aspects such as security, cost, performance, traceability of transactions made through the application, etc. Further, preferences over an attribute may depend on the preference over other attributes. The following is a set of sample user preferences[4] on the attributes of the software implementation.

p_1 Among implementations that are *Secure*, it is more important to have *Cheaper transaction costs* and *Performance*, rather than just *Traceability*.

p_2 Among implementations that do not satisfy the *Traceability* non-functional property (NFP), *Security* is more important than *Cheaper transaction costs*.

Any design that satisfies all four aspects is trivially the most preferred design, and will be consistent with both the above preferences of the architect. However, note that such an implementation that simultaneously optimizes all non-functional aspects may not be realizable due to other constraints (e.g., two software components may be incompatible); then the question of interest is—what is the most preferred realizable implementation?

1.4   COUNTERMEASURES FOR NETWORK SECURITY

Consider a scenario where a system administrator (defender) comes to know of certain vulnerabilities in the network he is managing. The defender would like to deploy the set of countermeasures that best mitigate the security concerns while not rendering the system unusable.

Determining the preferred sets of countermeasures requires accounting for the preferences of the defender over various countermeasures that could be deployed in response to certain security concerns. Suppose that the countermeasure options considered by the defender are: setting up a firewall (FW), restricting the access privileges on the server (AP), increasing logging levels (LO), and stopping the service all together (ST). In this example, deploying fewer countermeasures is preferred to deploying more countermeasures, as each countermeasure incurs cost of deployment and also affects the system functionality as well as usability. An example set of preferences is given below.

---

[4]See [68] for more details on this example.

p_1 If FW is part of the defense strategy anyway, then AP is preferred to ST.

p_2 If AP is included in the strategy but LO is not considered, then ST is preferred to FW.

p_3 If LO is part of the strategy but not ST, then AP is preferred to FW.

p_4 If LO is not part of the strategy, then including both FW and AP is preferred to ST.

p_5 If ST is not considered, then including LO along with FW is preferred to AP.

The objective of the defender would be to obtain an ordering of countermeasure sets starting from the most preferred one, so that he can identify the most preferred set of countermeasures that satisfies the security concerns.

1.5   MINIMIZING CREDENTIAL DISCLOSURE

Consider a client who is interested in obtaining some financial quote (e.g., auto and/or home insurance, mortgage, etc.) using an online service. Suppose that there are multiple servers that provide the required service, and each server's access control policy requires a combination of several credentials from the client before granting access to the service. The client may want to obtain the quote from the service that requires the least sensitive credential information. Here, the sensitivity of credentials as viewed by the client induces preferences over services that require various subsets of these credentials.

Consider four such credentials: the client's name, residential address, bank account number, and bank routing number. The client may have qualitative preferences over the relative importance of his credentials based on their sensitivity. The rationale behind these preferences is that the client would like to make it impossible (or at least difficult) for a third party to perform any financial transaction maliciously posing as the client. Therefore, from the client's perspective, the objective is to choose the server that provides the desired financial service by requiring the least sensitive set of client credentials. Consider the following qualitative preferences specified by the client:

p_1 If my bank account number is disclosed to the server, I would rather give my address than my bank's routing number to the server. This is because my bank account number along with the bank routing number identifies my bank account precisely, and hence it is highly sensitive information compared to my bank account number and address.

p_2 If I have to disclose my address without having to disclose my name, then I would prefer giving my bank's routing number over my bank account number. However, this preference does not hold when I have to disclose my name along with my address, because the combination of my name, address, and bank routing number is not any less sensitive than my name, address, and bank account number. In both cases, a malicious party needs to guess one of the credentials—bank account number or bank routing number—to gain access to important financial information.

p_3 Because I would like to protect as many details as possible regarding my bank account, when I don't have to disclose my bank account number I would provide my name and address rather than my bank's routing number.

Based on these preferences, the client may be interested in finding successively more sensitive sets of credentials (starting from the empty set) and verify whether a set of credentials is sufficient to satisfy the access control policy of a server providing the desired service. The idea is that any server that accepts this least sensitive acceptable set of credentials may be selected to provide the service to the client, as it poses the least risk of compromise to privacy. The client may also be interested in knowing whether one set of credentials is riskier than another set to disclose to a server.

2   ORGANIZATION OF THE BOOK

The preceding examples motivate the need for effective techniques for representing and reasoning about qualitative preferences. In the last decade, several formal languages have been developed for representing qualitative preferences, such as CP-nets [16], TCP-nets [18], CP-Theories [94], CI-nets [17]. However, an obstacle to using these languages for real-world applications is that reasoning in these languages is generally hard. For example, testing whether one alternative is preferred to another with respect to even a simple language, namely CP-nets, is PSPACE-complete [41]. Against this background, this book presents a suite of modern approaches to reasoning with qualitative preferences that leverage recent advances in formal methods and model checking [27, 76].

The book is organized as follows:

Chapter 2. This chapter reviews popular qualitative preference languages such as CP-nets, TCP-nets, CI-nets, and CP-theories and gives their syntax and a well-studied semantics called *ceteris paribus*[5] semantics in a unified notation. The chapter introduces two key preference reasoning tasks, namely, dominance testing and consistency testing. The chapter introduces a key data structure for representing the semantics of qualitative preferences, namely the *induced preference graph*. The chapter also discusses how preferences in the examples provided in Chapter 1 can be formalized using these languages.

Chapter 3. This chapter provides an overview of model checking, which is a formal verification technique that has been successfully used in several applications. Model checking forms the basis of the approaches presented in all the following chapters, and so this chapter is a prerequisite for the rest of the book, except for the last chapter. This chapter presents the syntax and semantics of a temporal logic that we will use in subsequent chapters, and describes a data structure called the *Kripke structure* used to succinctly express the semantics of temporal models. Those familiar with temporal logic model checking and CTL can skip this chapter.

---

[5]*Ceteris paribus* is a Latin term for "all else being equal."

Chapter 4. This chapter first presents a way to deal with the intractability of preference reasoning tasks by restricting the expressivity of preferences. The chapter studies a language in which preferences are always unconditional, provides its syntax and semantics, and contrasts the semantics and expressivity of this language with other conditional preference languages. The chapter shows how dominance testing becomes tractable in polynomial time by the proposed restriction, and mentions applications where such a preference language might be appropriate.

The second part of this chapter outlines a way to translate the induced preference graph (which represents the semantics of a set of preferences) into a Kripke structure (which succinctly stores the semantics of a temporal model) in a way that the preference semantics is preserved in the Kripke structure. The latter part of the chapter also describes how one can translate preference queries with respect to a set of preferences into an equivalent set of formulas in the temporal logic CTL, such that the CTL formula is satisfied whenever the preference query holds. The chapter also describes a method to use the model checker to (whenever possible) obtain a justification of why a preference query is satisfied (or not) with respect to a set of preferences.

Chapter 5. This chapter defines the notions of the equivalence of two sets of preferences, and what it means for one set of preferences to subsume another, along with their formal semantics. This problem is interesting in settings where it is necessary to reason about the preferences of multiple agents. The chapter proceeds to extend the technique presented in Chapter 4 in order to automatically verify preference equivalence and preference subsumption with respect to two sets of preferences, and if the result is negative, then how to extract an evidence justifying the result.

Chapter 6. This chapter extends the model checking techniques presented in Chapters 4 and 5 to obtain an ordering of alternatives starting from the most preferred alternatives to the least preferred. The chapter also provides a uniform way of handling inconsistencies in preferences, i.e., cyclic preferences over alternatives (e.g., A is preferred to B, which is in turn preferred to A). The chapter also characterizes the ordering computed by the method presented as a useful extension of the original preference order induced by the stated preferences.

Chapter 7. This chapter describes a preference reasoner tool called CRISNER [80] that implements the model checking approach to qualitative preference reasoning in CP-nets, TCP-nets, and CP-theories. The tool's architecture, XML-based input and output formats, and some details regarding realizing the techniques in this book using the NuSMV model checker are discussed. The chapter concludes with a set of projects one can work on, to extend the tool to support other preference languages and reasoning tasks.

Chapter 8. The last chapter summarizes the book's contents and presents directions for future work extending the techniques for preference reasoning presented.

Appendices. The source code for the Kripke structures models translated from the example preference specifications, sample CTL queries corresponding to preference reasoning tasks along with the respective traces from the NuSMV model checker, and sample XML files used as input for the preference reasoning tool presented in Chapter 7 are provided as reference for the readers. This will enable even those who do not have access to the model checker to follow how the techniques work for the examples in the book.

Readers will find it useful to follow one of the following study plans based on their own backgrounds and interests.

- The reader who is unfamiliar with qualitative preference languages, their syntax, and semantics is advised to begin from Chapter 2 and proceed chapter by chapter up to the end of the book for a complete study of all topics.

- Assuming that the reader is already familiar with the syntax of the qualitative preference languages and their *ceteris paribus* semantics, one can begin with Chapter 4 and proceed chapter-wise up to the end of the book for a detailed study of the model checking approach to preference reasoning in these languages.

- For a reader who is interested only in an overview of the qualitative preference languages, and is more eager to learn about trading off expresssivity for tractability of reasoning, it is best to study Chapter 2 and move on to Chapter 4, particularly Section 1.

The reader is encouraged to refer to the appendices containing source code for many examples discussed in the book, and also to try out some of the SMV modeling and CTL verification exercises for a better understanding of the subject matter.

CHAPTER 2

Qualitative Preference Languages

In the previous chapter, we reviewed some real-world applications where preferences play a vital role in choosing the most desirable alternative from among a set of candidates. User preferences over the alternatives were expressed as a qualitative (relative) ordering over the set of attributes and their respective values. This chapter aims to review some of the key formalisms for representing and reasoning with such qualitative preferences. We begin the chapter by introducing qualitative preferences as binary relations over the set of alternatives. We review a broad class of qualitative preference languages that allow users to express various kinds of preference relations over attributes and their domains, and their semantics in terms of an induced preference graph that encodes a partial order over set of alternatives with respect to a given set of preferences. We will then define a set of reasoning tasks, e.g., checking whether one alternative dominates another with respect to a given set of preferences. For a thorough review of the literature on qualitative preference languages, the interested reader is encouraged to refer to [35].

1   PRELIMINARIES

Our aim is to represent and reason about the relative preference of outcomes or alternatives with respect to a set of the user's qualitative preferences.[1] We assume that in the general case, an alternative in a decision problem may be described in terms of a set of *attributes* or *preference variables* (or simply *variables*) that the user cares about when making his choice. In the credentials example in Section 1.5, the various subsets of user credentials form the set of alternatives; in the cyberdefense example in Section 1.1 each alternative is described in terms of a valuation to the three variables describing a vulnerability. Formally, we use the following notations for the concepts just described.

1.1   NOTATION

Let $X = \{x_i \mid 0 < i \leq n\}$ be a set of preference variables or attributes. For each $x_i \in X$ let D_i be the set of possible values (i.e., domain) such that $x_i = v_i \in D_i$ is a valid assignment to the variable x_i. We use Φ, Ψ, and Ω (indexed, subscripted, or superscripted as necessary) to denote subsets of X. The set $\mathcal{O} = \{\prod_{x_i \in X} D_i\}$ of assignments to variables in X is called the set of all

---

[1]In the rest of the book, we will use the term "preference" for "qualitative preference," unless otherwise mentioned.

possible *alternatives*. The *valuation* of an alternative $\gamma \in \mathcal{O}$ *with respect to a variable* $x_i \in X$ is denoted by $\gamma(x_i) \in D_i$. The set of assignments with respect to a subset $\Phi = \{x_{j_1} \ldots x_{j_m}\} \subseteq X$ is given by $\mathcal{O}_\Phi = \{\gamma \mid \prod_{x_i \in \Phi} D_i\}$, where γ is an assignment to the variables in Φ, denoted by the tuple $\gamma := \langle v_1, v_2, \ldots, v_m \rangle$, such that $v_i = \gamma(x_i) \in D_i$ for each $x_i \in \Phi$.

Example 2.1   For the cyberdefense example in Section 1.1, the vulnerability that has a freely available exploit code and a Fix-Available with a high access complexity is denoted by the tuple $\langle E = $ Code-Available, $A = $ High, $F = $ Fix-Available$\rangle$, or simply $\langle$Code-Available, High, Fix-Available$\rangle$ when the attributes can be inferred from the context.

Note that in a CI-net (see example in Section 1.3), the meaning of "attribute" is different from what it means in the context of a CP-net or TCP-net or CP-Theory. When talking about CI-nets, an attribute is a boolean property (binary variable), and is "true" in an alternative if it is "satisfied" by the alternative. Each alternative is described by the set of attributes that are included in (or properties satisfied by) it, and the preferences over alternatives are specified in terms of the presence or absence of attributes in an alternative. To represent such alternatives uniformly in terms of tuples of valuations of attributes as above, we will use a boolean variable for each attribute, whose truth valuation denotes the inclusion of the attribute in the alternative.

Example 2.2   We denote an alternative that includes the attributes "Cheaper transaction costs" (C) and "Performance" (R) (but not "Security" (S) and "Traceability" (T)) in the example in Section 1.3 by

$$\langle C = \texttt{true}, R = \texttt{true}, S = \texttt{false}, T = \texttt{false} \rangle$$

As a notational convenience, in such settings we will denote an alternative that includes the set $\Phi \subseteq X$ of attributes, simply by the subset Φ. For example, the above alternative will be represented as the set $\{C, R\}$; the alternative that includes only "Security" and "Cheaper transaction costs" is denoted as $\{C, S\}$ (rather than $\langle C = \texttt{true}, R = \texttt{false}, S = \texttt{true}, T = \texttt{false} \rangle$); and the preference (p_1) represents the preference of the alternative $\{C, R, S\}$ over the alternative $\{S, T\}$.

1.2   SUCCINCT PREFERENCE SPECIFICATION

Given a set $\mathcal{O}$ of n alternatives, a direct specification of a binary preference relation $\succ$ over $\mathcal{O}$ is difficult, as it requires the user to compare up to $O(n^2)$ pairs of alternatives, which is prohibitive in time because n is exponentially large. Hence, many preference languages allow for succinct specification of the preference relation over alternatives in terms of *preference relations over the set of attributes that describe the alternatives and their respective valuations (i.e., domains).*

Preference Relations   Qualitative preference relations can be either (a) **intra-variable preference relations over valuations** of an attribute; or (b) **relative importance preference relations over attributes**. Note the subtle distinction between relative importance and intra-attribute preferences—while relative importance specifies a preference relation over the set of attributes, individual intra-variable preferences with respect to each variable specify preference relations over the set of valuations of that variable. A language that allows for specification of relative importance preferences independent of the specification of the intra-variable preferences may enable succinct representation of user preferences; however, it necessitates the development of techniques for reasoning about preferences over alternatives that combine both types of user preferences in a suitable way.

For any $\Phi \subseteq X$, we will use the notation $\succ_\Phi$ to denote a preference relation over D_Φ, the set of partial assignments to attributes in Φ. For a single attribute $x_i \in X$, the intra-attribute preference relation over its valuations (D_i) will be denoted by $\succ_{\{x_i\}}$ or alternatively $\succ_i$. For example, to formally specify that the valuation v_i is preferred to the valuation v_i' for attribute x_i where $v_i, v_i' \in D_i$, we will write $x_i = v_i \succ_i x_i = v_i'$. We will use the notation $\triangleright$ to indicate relative importance between attributes or between sets of attributes.

Example 2.3   Intra-attribute Preference   In the example in Section 1.1, that *Low* access complexity is preferred to *High* is an unconditional intra-attribute preference:

$$A = \text{Low} \succ_A A = \text{High}$$

In the same example, that a fix to the vulnerability is unavailable is preferred to the case when a Fix-Available is available, is a conditional intra-attribute preference:

$$E = \text{Code-Available} \Rightarrow F = \text{Fix-Unavailable} \succ_F F = \text{Fix-Available}$$

Example 2.4   Relative Importance   That A is more important than E in the TCP-net in Section 1.1 is an unconditional relative importance preference: $A \triangleright E$. In the same example with a CP-theory, that A is more important than the set of attributes $\{E, F\}$ is an example of One-Many relative importance, i.e., $\{A\} \triangleright \{E, F\}$.

Further, the preference p_1 in the Section 1.3 states conditional Many-One relative importance in a CI-net: among two sets of attributes that include S, $\{C, R\} \triangleright \{T\}$ meaning that one that includes both C and R is preferred to one that includes only T.

Preference Statements   In all the above languages, preferences are expressed in terms of a set $P = \{p_i\}$ of preference statements. Each statement p either specifies a binary relation over the domain of a particular variable (intra-variable preference) or a binary relation over the set X of preference variables (relative importance). The semantics of the preference languages define how these succinct statements of preference are interpreted in order to determine the preference over alternatives. Several different preference semantics may be defined for a given preference language depending on how the stated preferences over the variables are interpreted and translated into preferences over alternatives; we focus on the most popular interpretation, namely the *ceteris paribus* semantics. We will first describe the representation scheme of CP-nets, TCP-nets, CP-Theories, and CI-nets, followed by their semantics.

2   QUALITATIVE PREFERENCE LANGUAGES

We consider members of the *preference network* family of languages that are popular, including conditional preference networks (CP-nets) [16], generalized CP-nets (GCP-nets) [41], and trade-off enhanced conditional preference networks (TCP-nets) [18]. The above languages allow user preferences to be modeled using an intuitive graphical representation scheme. We also consider other qualitative preference languages such as conditional preference theories (CP-Theories) [94–96], and conditional importance networks (CI-nets) [17] that do not have a simple graphical representation scheme, but allow preferences to be modeled in terms of a set of preference statements. We do not consider UCP-nets [13], an extension of CP-nets that allow the representation of quantitative preferences (utility information) rather than simple qualitative preference orderings.

2.1   REPRESENTING QUALITATIVE PREFERENCES

The need for representing such qualitative preferences in a succinct, user-friendly way has been addressed by several authors in the past. Of particular relevance are *conditional preference networks* (CP-nets[16] and GCP-nets[41]) that allow the user to express a set of conditional preferences over the domain of an attribute of the form "the valuation x_1 is preferred to the valuation x_1' for attribute X_1 under a given condition" (e.g., p_1–p_3 in the cyberdefense example of Section 1.1). TCP-nets [18] further allow simple conditional relative importance preferences over pairs of attributes of the form X_1 is more important than X_2 (e.g., p_5). The above languages also support preference input using a compact graphical representation scheme. Some other related languages such as CI-nets and CP-Theories allow statements of relative importance among *sets* of attributes. In particular, CP-Theories [96] allow statements of the form "X_1 is more important than $\{X_2, X_3\}$" (e.g., p_6). CI-nets [17] allow unconditional and monotonic intra-attribute preferences, and conditional relative importance preferences over sets of attributes of the form "$\{X_1, X_2\}$ is more important than $\{X_3, X_4\}$" (e.g., p_7 can be expressed in a CI-net but not in the other languages). The above preference languages allow the succinct expression of user preferences over alternatives in terms of a **preference specification** P, which is a set of preference statements p over the preference variables V. The formal syntax and interpretation, as well as compact graphical representation schemes for preference specifications in these languages, are described in detail later.

There are also other graphical models for representing qualitative preferences, notably preference trees (P-trees) and lexicographic preference trees (LP-trees) [11, 58, 59], as well as logic-based models[2] that we do not discuss in this book.

Table 2.1: Qualitative preference languages and the types of preference relations they allow to express. The first column describes the kind of preference relations that are supported (or not supported) by the different preference languages. A checkmark (✓) indicates that the language allows the corresponding preference relation to be expressed. * A CI-net allows only monotonic intra-variable preference.

Preference Relation Type	CP-net	TCP-net	CP-Theory	CI-net
Unconditional Intra-variable Preference	✓	✓	✓	✓*
Unconditional Relative Importance		✓	✓	✓
Conditional Intra-variable Preference	✓	✓	✓	
Conditional Relative Importance		✓	✓	✓
One-Many Relative Importance			✓	✓
Many-Many Relative Importance				✓

2.2   PREFERENCE SEMANTICS

The semantics of the preference languages define how these succinct statements of preference over individual variables may be interpreted in order to determine the preference over alternatives.

A well-studied semantics for interpreting the user's statements of preference in a preference network is the *ceteris paribus* or "all else being equal" semantics.

- The *ceteris paribus* interpretation of an *intra-attribute preference* statement over the domain of an attribute X rules that an alternative α is preferred to another β if and only if all other attributes being equal, the assignment to X in α is preferred to the assignment to X in β.

- A *relative importance* statement of the form X is more important than Y is interpreted as follows: An alternative is preferred to another if and only if it has preferred values for the more important attribute X, all other attributes *except* Y being equal.

---

[2]For a more elaborate discussion on these other preference languages, please refer to [52].

The second kind of (relative importance) preference is supported by TCP-nets, CI-nets, and CP-theories. More expressive relative importance statements can be represented in CP-theories and CI-nets:

- The statement "an attribute X over a set $\mathcal{Y}$ of other attributes" is interpreted as follows. An alternative is preferred to another if and only if it has preferred values for the more important attribute X, all other attributes *except those in* $\mathcal{Y}$ being equal.

The *ceteris paribus* semantics for a set of preference statements is given by an **induced preference graph** that is induced from the intra-variable and relative importance preference statements. The nodes of this induced graph correspond to the set of all alternatives (i.e., the set of all assignments to all preference variables). Each edge of the induced preference graph is directed from a less preferred to a more preferred alternative, and computing the *ceteris paribus* semantics amounts to making traversals on the induced preference graph. In terms of the induced preference graph, an alternative **dominates** another if and only if there exists a path from the less preferred to the more preferred alternative.

According to the *ceteris paribus* interpretation [15, 45] of preferences, each preference statement $p \in P$ allows a set of changes to the valuation(s) of one or more variables in an alternative β in order to obtain a more preferred alternative α, while other variables remain fixed. Such a change is called an **improving flip**.

Definition 2.5 Improving flip.   Given a preference specification P, there is said to be an improving flip from an outcome β to another outcome α if and only if there is some preference statement $p \in P$ that induces the dominance of α over β according to the interpretation of the intra-variable and relative importance preference statements in P, as given by the semantics of the language of P.

Example 2.6 Improving flip   For instance in Section 1.1, the alternative $\beta = \langle E = $ Code-Unavailable, $A = $ Low, $F = $ Fix-Unavailable$\rangle$ can be changed to obtain a more preferred alternative $\alpha = \langle E = $ Code-Available, $A = $ Low, $F = $ Fix-Unavailable$\rangle$ by modifying the valuation of E in β using preference statement p_1. In other words, p_1 **induces** an improving flip from β to α.

The semantics of a preference specification P in any of the above languages (namely, CP-nets, TCP-nets, CP-Theories, and CI-nets) is given in terms of a (strict) partial order $\succ$ over the alternatives, where $\alpha \succ \beta$ if and only if there is a sequence of alternatives $\beta = \gamma_1, \gamma_2, \ldots \gamma_n = \alpha$ such that for all $1 \leq i \leq n$, there is an improving flip from γ_i to γ_{i+1} with respect to some $p \in P$. Such a sequence is called an **improving flipping sequence** from β to α.

The only difference in the semantics of CP-nets, TCP-nets, CP-Theories, and CI-nets is the definition of what constitutes a valid improving flip. For example, CP-nets can represent only intra-variable preferences over the valuations of one variable at a time, whereas TCP-nets can additionally represent relative importance between a pair of variables at a time, and CP-Theories allow relative importance of one variable over a set of variables at a time. As a result, the improving flips for CP-nets allow at most one variable to change at a time; those for TCP-nets allow at most two variables to be changed at a time; and those for CP-Theories allow multiple variables to be changed at a time.

We now describe the representation scheme of CP-nets, TCP-nets, CP-Theories, and CI-nets and their improving flipping sequence semantics.

2.3   CP-NETS

CP-nets [16] use a compact graphical model to specify the conditional intra-variable preferences $\succ_{\{x_i\}}$ individually for each variable x_i. Each node in the graphical representation of a CP-net C corresponds to a variable $x_i \in X$, and each *dependency* edge (x_i, x_j) in the graph captures the fact that the intra-variable preference $\succ_{\{x_j\}}$ with respect to variable x_j is dependent (or conditioned) on the valuation of x_i. For any variable x_j, the variables $\{x_i : (x_i, x_j)$ is an edge$\}$ are called the *parent* variables, denoted $\rho(x_j)$. Each node x_i in the graph is associated with a *conditional preference table* that maps each possible assignment to the parents $\rho(x_i)$ to a total order over D_i. An *acyclic* CP-net is one that does not contain any cycles in the dependency graph. GCP-nets [41] generalize CP-nets with binary variables in that they (a) allow conditional preference tables to be partially defined, i.e., the user may choose to specify neither $v_i \succ_{\{x_i\}} v_i'$ nor $v_i' \succ_{\{x_i\}} v_i$ for some assignment to the parents $\rho(x_i)$; and (b) do not require the assignments to the parents $\rho(x_i)$ in the conditional preference table of x_i to be exhaustive, i.e., the mapping from the set of assignments of $\rho(x_i)$ to partial orders over values of x_i can be partial. CP-nets and GCP-nets can be further generalized such that (a) each variable x_i has an arbitrary domain D_i; (b) conditional preference table of each variable x_i provides a *partial* (not necessarily complete) mapping from some assignments of $\rho(x_i)$ to conditional preference relations $\succ_{\{x_i\}}$ that are strict partial (not necessarily total) orders over D_i. In the rest of the book, we will use "CP-nets" to refer to such a generalization of CP-nets and GCP-nets.

The preferences encoded in a CP-net can be equivalently expressed in terms of a set of preference statements of the form:

$$\varrho : x_i = v_i \succ_{\{x_i\}} x_i = v_i'$$

where $\varrho \in D_{\rho(x_i)} \subseteq X$, $x_i \in X \setminus \Phi$ and $v_i, v_i' \in D_i$. When $\varrho = $ true, we simply write $x_i = v_i \succ_{\{x_i\}} x_i = v_i'$. Then, each preference ordering $\succ_{\{x_i\}}$ of D_i under the condition ϱ corresponds to a row in the conditional preference table of x_i in the CP-net.

Example 2.7   Figure 2.1(a) shows a CP-net over the set $\{E, A, F\}$ of variables for the cyberdefense example (see Section 1.1). The conditional preference tables for each variable is shown as a box next to the variable. Note that the conditional preference table of F is incomplete, i.e., the ordering over the domain of F is partial when E takes the valuation Code-Unavailable, and it is an acyclic CP-net because there is only one dependency edge from E to F. This CP-net can

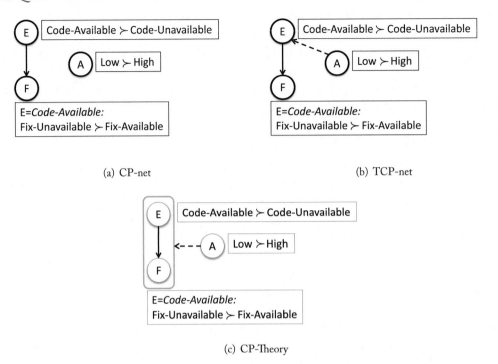

(a) CP-net (b) TCP-net

(c) CP-Theory

Figure 2.1: Risk priorities of a cyberdefender.

be equivalently expressed in terms of a preference specification P^{CP} consisting of the following preference statements. An illustration of the actual partial order on the alternatives induced by this CP-net will be given in Chapter 4.

p_1. $E = \text{Code-Available} \succ_E E = \text{Code-Unavailable}$

p_2. $A = \text{Low} \succ_A A = \text{High}$

p_3. $E = \text{Code-Available} : F = \text{Fix-Unavailable} \succ_F F = \text{Fix-Available}$

2.4 TCP-NETS

TCP-nets [18] allow the users to specify *pairwise* relative importance among variables in addition to conditional intra-variable preferences as in CP-nets. As in CP-nets, each node in the TCP-net corresponds to a variable x_i and is associated with a conditional preference table that (possibly partially[3]) maps assignments to $\rho(x_i)$ to conditional preference relations $\succ_{\{x_i\}}$ that are strict partial orders over D_i. The relative importance of a variable x_i over x_j is denoted by $x_i \rhd x_j$, and

---

[3]The original definition of TCP-nets [18] requires this mapping to be total; however, we relax this condition, as the reasoning techniques we develop later in the book can be applied to the more general definition.

represented as another type of edge (x_i, x_j) in the TCP-net graph, differently from the conditional dependency edges. Each relative importance edge could be either *unconditional* (directed edge) or conditioned on a set of *selector* variables (analogous to parent variables in the case of intra-variable preferences). Each edge (x_i, x_j) describing *conditional* relative importance is undirected (because whether or not one variable is more important than the other could depend on the conditions) and is associated with a conditional importance table (CIT, analogous to the conditional preference table), mapping an assignment of the selector variables to either $x_i \rhd x_j$ or vice versa. Note that this mapping may be partial, which means that the relative importance relation between x_i and x_j may be specified only for certain values of the set of selector variables.

A TCP-net can be equivalently expressed in terms of a set of preference statements of the form:

$$\varrho : x_i = v_i \succ_{\{x_i\}} x_i = v_i' \ [\Omega]$$

where $\{x_i\}$, $\Phi = \rho(x_i)$ and Ω are disjoint subsets of X, $v_i, v_i' \in D_i$, $\varrho \in D_\Phi$ and $|\Omega| = 0$ or 1.[4] Specifically, the notation $[\Omega]$ at the end of the above preference statement denotes the relative importance of the variable x_i over $x_j \in \Omega$ if $\Omega = \{x_j\}$ (under the condition ϱ), and when $\Omega = \emptyset$, the statement expresses the conditional intra-attribute preferences for x_i. When $\Phi = \emptyset$, x_i is unconditionally more important than x_j, and hence the corresponding relative importance edge between x_i and x_j is directed from x_i to x_j; and when $\Phi \neq \emptyset$, the relative importance edge between x_i and x_j is undirected (to allow different importance relationships under different conditions).

Example 2.8 Figure 2.1(b) shows a TCP-net over the set $\{E, A, F\}$ of variables for the cyberdefense example (see Section 1.1). The preference specification P^{TCP} representing this TCP-net will include the preference statements p_1 and p_3 of the specification P^{CP} in Example 2.7 as the intra-attribute preferences are common. However, the preference over valuations of A will be represented differently in P^{TCP}, due to the relative importance of A over E (dotted edge from A to E); hence the preference statement p_2' (which further elaborates on p_2 in Section 1.1) in P^{TCP} will be as follows.

p_2'. $A = \text{Low} \succ_A A = \text{High} \ [\{E\}]$

2.5 CP-THEORIES

CP-Theories [96] extend TCP-nets by further allowing the specification of the relative importance of one variable over a *set* of variables conditioned on another set of variables. A CP-Theory consists of statements of the form

$$\varrho : x_i = v_i \succ_{\{x_i\}} x_i = v_i' \ [\Omega]$$

---

[4]Although Ω is either an empty set or a singleton, we still choose to represent it as a subset of variables, as it makes it easier to describe the next language, namely CP-theories, that allows for subsets of variables with more than one variable in Ω.

where ϱ is an assignment to the set $\Phi \subseteq X$ of variables that defines the condition under which the preference holds, $v_i, v_i' \in D_i$, $\Omega \subseteq X$, and $\Phi, \Omega, \{x_i\}$ and $(X - \Phi - \Omega - \{x_i\})$ are disjoint. Specifically, the notation $[\Omega]$ at the end of the above preference statement denotes the relative importance of the variable x_i over the set Ω of variables under the condition ϱ. Note that CP-nets can be expressed as CP-Theories by fixing $\Omega = \emptyset$ (i.e., $|\Omega| = 0$); and TCP-nets can be expressed as CP-Theories by fixing $|\Omega| = 0$ or 1. In particular, suppose that a TCP-net contains the statement $x_i \triangleright x_j$, and the intra-attribute preferences of x_i are defined by $\succ_{\{x_i\}}$, then the same can be represented in terms of a set of CP-Theory statements such that for each statement $\varrho : x_i = v_i \succ_{\{x_i\}} x_i = v_i'$ in the TCP-net, there is a new CP-Theory statement $\varrho : x_i = v_i \succ_{\{x_i\}} x_i = v_i' [\{x_j\}]$.

Example 2.9 Figure 2.1(c) shows a CP-theory over the set $\{E, A, F\}$ of variables for the cyberdefense example (see Section 1.1). The preference specification P^{CPT} representing this CP-theory will include the preference statements p_1 and p_3 as in the specifications P^{CP} and P^{TCP}; and the preference over valuations of A will be represented in P^{CPT} differently from P^{TCP} due to difference in the statement of relative importance of A over the set $\{E, F\}$ (illustrated as dotted edge in Figure 2.1(c)). Hence the preference statement p_2'' (which further elaborates on p_2 in Section 1.1) in P^{CPT} will be as follows:

p_2''. $A = \text{Low} \succ_A A = \text{High} [\{E, F\}]$

2.6 CI-NETS

CI-nets [17] allow for the specification of more expressive relative importance preferences as compared to TCP-nets, namely, conditional relative importance over arbitrary sets of variables (as opposed to relative importance over just singletons). However, CI-nets place two restrictions on the intra-variable preferences: (a) the domains of the variables are binary, i.e., $\forall x_i \in X : D_i = \{\text{true}, \text{false}\}$; and (b) the intra-variable preferences are unconditional and monotonic, i.e., $\forall x_i \in X : \text{true} \succ_{\{x_i\}} \text{false}$.[5] As a result, an alternative γ corresponds to a subset of X, and the set of all possible alternatives $\mathcal{O}$ corresponds to the powerset of X. A CI-net consists of statements of the form: $\Phi^+, \Phi^- : \Psi_1 \succ \Psi_2$, which states that "among two alternatives that include variables in Φ^+ and exclude variables in Φ^-, the alternative that includes those in Ψ_1 is preferred to one that includes those in Ψ_2." Such set-based relative importance cannot be expressed in CP-nets, TCP-nets, or CP-theories. On the other hand, CI-nets restrict intra-variable preferences to be unconditional and monotonic, and hence they cannot express conditional intra-variable preferences which is possible in TCP-nets and CP-nets.

Note that in CP-nets, TCP-nets, and CP-Theories, each preference statement in the language can be represented in terms of preferences over the valuations of a single attribute x_i (and

---

[5]This order can be domain specific [68, 69].

possibly the relative importance of that attribute over others by specifying a non-empty Ω). However, in the case of CI-nets such a representation is not always possible because each preference statement may define the relative importance of a set (Ψ_1) of attributes over another (Ψ_2). Hence, the above CI-net statement can be translated into the following preference statement:

$$\varrho : (\bigwedge_{x_i \in \Psi_1} x_i = \texttt{true} \wedge \bigwedge_{x_j \in \Psi_2} x_j = \texttt{false}) \quad \succ_{\Psi_1 \cup \Psi_2} \quad (\bigwedge_{x_i \in \Psi_1} x_i = \texttt{false} \wedge \bigwedge_{x_j \in \Psi_2} x_j = \texttt{true})$$

where ϱ is the partial assignment to $\Phi^+ \cup \Phi^-$, such that $\forall x_i \in \Phi^+ : \varrho(x_i) = \texttt{true}$ and $\forall x_j \in \Phi^- : \varrho(x_j) = \texttt{false}$. In the above, the preference relation $\succ_{\Psi_1 \cup \Psi_2}$ is over $D_{\Psi_1 \cup \Psi_2}$.

Example 2.10   The CI-net described in Example 1.3 is expressed in the following form.

p_1. $\{S\}, \{\} : \{C, R\} \succ \{T\}$

p_2. $\{\}, \{T\} : \{S\} \succ \{C\}$

 The above CI-net can be equivalently expressed in terms of the preference specification P^{CI} consisting of the following preference statements.

p_1. $S = 1 : (C = 1 \wedge R = 1 \wedge T = 0) \succ (C = 0 \wedge R = 0 \wedge T = 1)$

p_2. $T = 0 : (S = 1 \wedge C = 0) \succ (S = 0 \wedge C = 1)$

Example 2.11   For the CI-net described in the countermeasure example in Section 1.4, let $a :=$ "Increase Logging Level," $b :=$ "Restrict Access Privilege," $c :=$ "Stop Service," and $d :=$ "Setup a Firewall." The following statements then constitute the preference specification in that example.

p_1. $\{d\}, \{\} : \{b\} \succ \{c\}$

p_2. $\{b\}, \{a\} : \{c\} \succ \{d\}$

p_3. $\{a\}, \{c\} : \{b\} \succ \{d\}$

p_4. $\{\}, \{a\} : \{b, d\} \succ \{c\}$

p_5. $\{\}, \{c\} : \{a, d\} \succ \{b\}$

The formal representation of the above CI-net statements are as follows.

p_1. $d = 1 : (b = 1 \wedge c = 0) \succ (b = 0 \wedge c = 1)$

p_2. $(b = 1 \wedge a = 0) : (c = 1 \wedge d = 0) \succ (c = 0 \wedge d = 1)$

p_3. $(a = 1 \wedge c = 0) : (b = 1 \wedge d = 0) \succ (b = 0 \wedge d = 1)$

p_4. $a = 0 : (b = 1 \wedge d = 1 \wedge c = 0) \succ (b = 0 \wedge d = 0 \wedge c = 1)$

p_5. $c = 0 : (a = 1 \wedge d = 1 \wedge b = 0) \succ (a = 0 \wedge d = 0 \wedge b = 1)$

2.7   RELATIVE EXPRESSIVE POWER

In terms of expressive power, CP-nets is the least expressive formalism, allowing only intra-attribute preferences with conditional dependencies among variables, and no relative importance preferences. TCP-nets are more expressive than CP-nets [18], as they allow relative importance preferences over individual variables in addition to CP-net language. CP-theories are more expressive than TCP-nets [93, 94] because they more generally allow specification of intra-attribute preferences with conditional dependencies, as well as specification of relative importance of one variable over a set of variables at the same time. CI-nets allow relative importance preference of one set over another set that cannot be expressed in CP-nets, TCP-nets, or CP-theories. For example, the statement p_1 in the CI-net P^{CI} cannot be expressed in CP-nets, TCP-nets, or CP-theories. On the other hand, CI-nets restrict intra-variable preferences to be unconditional and monotonic [17], and hence they cannot express conditional intra-variable preferences which is possible in TCP-nets and CP-nets. For example, the statement p_3 in the CP-net P^{CP} cannot be expressed in a CI-net. To summarize, CP-theories are more expressive than TCP-nets, which are in turn more expressive than CP-nets; and CI-nets are neither more nor less expressive than the other three languages.

3   REASONING WITH QUALITATIVE PREFERENCES

The preference languages discussed above allow the succinct expression of user preferences over alternatives in terms of a set $P = \{p_i\}$ of preference statements over the preference variables X. The semantics of the preference languages define how these succinct statements of preferences over individual variables may be interpreted in order to determine the preference over alternatives. Several different preference semantics may be defined for a given preference language depending on how the stated preferences over the variables are interpreted and translated into preferences over alternatives.

3.1   CETERIS PARIBUS PREFERENCE SEMANTICS

The semantics of a preference specification P over X is given in terms of the semantics of each of its preference statements. Consider a preference statement $p \in P$ that specifies the preference of one valuation of a set $\Phi \subseteq X$ of variables over another, conditioned on some valuation ϱ of another set $\rho \subseteq X$ of variables. The *ceteris paribus*, or "all else being equal" interpretation [15, 45] of p, induces the preference of one alternative over another if and only if (a) both alternatives respect the condition on ρ, i.e., agree on the respective valuations of each variable in ρ, (b) both alternatives agree on the valuation of $X \setminus \Phi \setminus \rho$, the set of variables that *do not appear* in p,[6] and (c) the first alternative is preferred to the second with respect to their respective valuations of Φ as specified by the $\succ_\Phi$ preference relation in p. Such an induced preference is called an *improving*

---

[6] *Ceteris paribus* means "all else equal," which is captured simply by the condition (b); however in the context of preference reasoning, the *ceteris paribus* semantics requires the conditions (a) and (c) as well.

flip, i.e., the result of "flipping" only the variables in Φ from a less preferred assignment of Φ to a more preferred assignment of Φ according to the preference stated in p. In other words, an improving flip from one alternative to another induced by p corresponds to changing the valuation of a set of variables in one alternative in order to obtain a preferred valuation for that set according to the conditional preference specified in p.

Preference languages differ in the types of preference statements they allow in a preference specification, namely intra-attribute preferences (all languages), relative importance preferences over individual attributes (TCP-nets, CP-Theories, CI-nets), or relative importance over sets of attributes (CP-Theories and CI-nets). Hence, we first describe the semantics of each of these types of preference statements in terms of the improving flips they induce over pairs of alternatives, and then describe the semantics of a preference specification in any of the above languages in terms of the flips induced by its constituent preference statements.

Semantics for Intra-variable Preference Statements

Consider an intra-variable preference statement p of the form $\varrho : v_i \succ_{\{x_i\}} v_i'$, which can be specified in all the languages we consider. Given two alternatives $\alpha, \beta \in \mathcal{O}$, the *ceteris paribus* interpretation of the statement p induces an improving flip from β to α if α and β differ only in x_i, and $\alpha(x_i) = v_i$ and $\beta(x_i) = v_i'$. In other words for any statement p, the valuation of only one variable can be flipped at a time, and that to a more preferred valuation with respect to p, while other variables remain fixed.

Definition 2.12   Preference Semantics for Intra-attribute Preference [16].   Given an intra-attribute preference statement p in a preference specification P of the form $\varrho : x_i = v_i \succ_{\{x_i\}} x_i = v_i'$ where $\varrho \in D_\Phi$, $\Phi = \rho(x_i) \subseteq X$ and two alternatives $\alpha, \beta \in \mathcal{O}$, there is an improving flip from β to α in $\delta(P)$ induced by p if and only if

1. $\exists x_i \in X : \alpha(x_i) = v_i$ and $\beta(x_i) = v_i'$,

2. $\forall x_j \in \Phi : \alpha(x_j) = \beta(x_j) = \varrho(x_j)$, and

3. $\forall x_k \in X \setminus \{x_i\} \setminus \Phi : \alpha(x_k) = \beta(x_k)$.

In the above definition, the first condition arises from the intra-attribute preference statement x_i; the second condition enforces the condition ϱ in p that states that α and β should concur on the parent variables of x_i; and the third enforces the *ceteris paribus* condition that states that α and β should concur on all the other variables.

Example 2.13   In the CP-net for the cyberdefense example in Figure 2.1(a), the preference statement p_2 induces improving flips from α to β whenever $\alpha(A) = High$, $\beta(A) = Low$, $\alpha(E) = \beta(E)$, and $\alpha(F) = \beta(F)$. In other words, p_2 represents an intra-variable preference over A that induces the following four improving flips, with the valuations of E and F being *ceteris paribus*, i.e., fixed.

- ⟨Code-Unavailable, High, Fix-Unavailable⟩ to ⟨Code-Unavailable, Low, Fix-Unavailable⟩

- ⟨Code-Available, High, Fix-Unavailable⟩ to ⟨Code-Available, Low, Fix-Unavailable⟩

- ⟨Code-Unavailable, High, Fix-Available⟩ to ⟨Code-Unavailable, Low, Fix-Available⟩

- ⟨Code-Available, High, Fix-Available⟩ to ⟨Code-Available, Low, Fix-Available⟩

On the other hand, the preference statement p_3 induces only two flips with the valuations of A and E being *ceteris paribus*.

- ⟨Code-Available, High, Fix-Available⟩ to ⟨Code-Available, High, Fix-Unavailable⟩

- ⟨Code-Available, Low, Fix-Available⟩ to ⟨Code-Available, Low, Fix-Unavailable⟩

Because the conditional preference for F is partially specified in the conditional preference table, alternatives are not comparable when $E =$ Code-Unavailable.

Semantics for Statements of Relative Importance Preference over Attributes

TCP-nets and CP-Theories allow the specification of relative importance preference of one variable over one or more variable respectively. We will treat CI-nets that allow specification of relative importance preferences of a set of variables over another separately below, because CI-nets represent attributes that are either present or absent in a given set.

TCP-nets and CP-Theories   Coming back to TCP-nets and CP-Theories, note that multiple variables can change in the same improving flip because a statement of relative importance of one attribute over others means that the user is willing to improve the valuation of the more important attribute *at the expense* of worsening the less important attribute(s).

Definition 2.14 Preference Semantics for Relative Importance of one Attribute over a Set [18, 94].   Given a relative importance preference statement p in a preference specification P of the form $\varrho : x_i = v_i \succ_{\{x_i\}} x_i = v_i'$ $[\Omega]$ where $\varrho \in D_\Phi, \Phi \subseteq X$ and two alternatives $\alpha, \beta \in \mathcal{O}$, there is an improving flip from β to α in $\delta(P)$ induced by p if and only if

1. $\exists x_i \in X : \alpha(x_i) = v_i$ and $\beta(x_i) = v_i'$,

2. $\forall x_j \in \Phi : \alpha(x_j) = \beta(x_j) = \varrho(x_j)$, and

3. $\forall x_k \in X \setminus \{x_i\} \setminus \Omega \setminus \Phi : \alpha(x_k) = \beta(x_k)$.

In the above definition, the first condition arises from the preference statement p on x_i; the second condition enforces the condition ϱ in p; and the third enforces the *ceteris paribus* condition and allows for unrestricted changes to the attributes that are less important than x_i (in Ω) when this preference statement is applied.

CI-nets   CI-nets allow relative importance of one set of attributes over another. Recall that a CI-net statement p of the form $\Phi^+, \Phi^- : \Psi_1 \succ \Psi_2$ can be translated into the following preference statement:

$$\varrho : (\bigwedge_{x_i \in \Psi_1} x_i = \texttt{true} \wedge \bigwedge_{x_j \in \Psi_2} x_j = \texttt{false}) \succ_{\Psi_1 \cup \Psi_2} (\bigwedge_{x_i \in \Psi_1} x_i = \texttt{false} \wedge \bigwedge_{x_j \in \Psi_2} x_j = \texttt{true})$$

where ϱ is the partial assignment to $\Phi^+ \cup \Phi^-$, such that $\forall x_i \in \Phi^+ : \varrho(x_i) = \texttt{true}$ and $\forall x_j \in \Phi^- : \varrho(x_j) = \texttt{false}$. In the above, the preference relation $\succ_{\Psi_1 \cup \Psi_2}$ is over $D_{\Psi_1 \cup \Psi_2}$, and *ceteris paribus* condition is enforced on the set $X \setminus (\Phi^+ \cup \Phi^- \cup \Psi_1 \cup \Psi_2)$. Note that in the case of the other languages, each preference statement defined preferences $(\succ_{\{x_i\}})$ over the valuations of a single attribute x_i and possibly the relative importance of that attribute over others (Ω). However, in the case of CI-nets each preference statement may define the relative importance of a set (Ψ_1) of attributes over another (Ψ_2), and hence it is represented by a preference relation over the partial valuations of the set $\Psi_1 \cup \Psi_2$ of attributes. For example, the preference p_1 in Section 1.3 specifies the relative importance of $\{C, R\}$ over T when both alternatives include S. In other words, the alternative $\langle C = \texttt{true}, R = \texttt{true}, S = \texttt{true}, T = \texttt{false}\rangle$ is preferred to $\langle C = \texttt{false}, R = \texttt{false}, S = \texttt{true}, T = \texttt{true}\rangle$. The second preference p_2 in the example translates to the following preference statement.

$$T = \texttt{false} : S = \texttt{true} \wedge C = \texttt{false} \succ S = \texttt{false} \wedge C = \texttt{true}$$

with the attribute P (that does not figure in the statement) being *ceteris paribus*, i.e, the two alternatives between whom the preference statement induces a preference agree on the valuation of R. In other words, $\langle C = \texttt{false}, R = \texttt{true}, S = \texttt{true}, T = \texttt{false}\rangle$ is preferred to $\langle C = \texttt{true}, R = \texttt{true}, S = \texttt{false}, T = \texttt{false}\rangle$ and $\langle C = \texttt{false}, R = \texttt{false}, S = \texttt{true}, T = \texttt{false}\rangle$ is preferred to $\langle C = \texttt{true}, R = \texttt{false}, S = \texttt{false}, T = \texttt{false}\rangle$. In addition the above preferences, the CI-nets by default include monotonicity of intra-attribute preferences for each attribute. The semantics for CI-nets was defined by Bouveret et al. [17] in terms of the set notation as follows.

Definition 2.15   Preference Semantics for Relative Importance of a Set of Attributes over Another Set[17].   Given a relative importance preference statement p of the form $\Phi^+, \Phi^- : \Psi_1 \succ \Psi_2$, and two alternatives $\alpha, \beta \in \mathcal{O}$, there is an improving flip from β to α in $\delta(P)$ induced by p if and only if either

1. (*Monotonicity Flip*) $\alpha \supset \beta$; or

2. (*Importance Flip*)

    (a) $\alpha \supseteq \Phi^+, \beta \supseteq \Phi^+, \alpha \cap \Phi^- = \beta \cap \Phi^- = \emptyset$;

(b) $\alpha \supseteq \Psi_1, \beta \supseteq \Psi_2, \alpha \cap \Psi_2 = \beta \cap \Psi_1 = \emptyset$;

(c) if $\gamma = G^\Phi \setminus (\Phi^+ \cup \Phi^- \cup \Psi_1 \cup \Psi_2)$ then $\gamma \cap \Psi_1 = \gamma \cap \Psi_2$.

In the above definition, the condition (1) states that the set is always preferred to its subset. The condition (2) states that given two sets ϱ and ϱ' both containing the elements in Φ^+ and not containing those in Φ^-, and if ϱ contains the elements in Ψ_1 and ϱ' contains those in Ψ_2 then ϱ is preferred to ϱ', all others being equal (which is ensured by the condition (2.2)).

3.2 SEMANTICS FOR A PREFERENCE SPECIFICATION AS INDUCED PREFERENCE GRAPHS

The semantics of a preference specification in each of the preference languages CP-nets, TCP-nets, CP-Theories, and CI-nets is given in terms of an *induced preference graph* [16–18, 96] that captures the set of all improving flips induced by all its preference statements.

Definition 2.16 Induced Preference Graph. Let P be a preference specification consisting of the set $\{p_i\}$ of preference statements in any of the above languages over the set X of variables. The *induced preference graph* $\delta(P) = G(A, E)$ is constructed as follows. The nodes A correspond to the set of all possible alternatives, i.e., complete assignments to all variables in X. Each directed edge $(\beta, \alpha) \in E$ from alternative β to alternative α is an *improving flip* from β to α, denoting that α is preferred to β according to the interpretation of some preference statement p_i in P.

From the above definition, we observe that the only difference in the semantics of the various preference languages we consider is the set of improving flips that they induce between alternatives in the induced preference graph.

Example 2.17 The induced preference graphs for the preference specifications P^{CP}, P^{TCP}, P^{CPT}, and P^{CI} are given in Figures 2.2, 2.3, 2.4, and 2.5 respectively. In the case of P^{CP}, P^{TCP}, and P^{CPT}, each improving flip (edge) that is induced by an intra-variable preference statement (annotated by the preference statement number) is shown by a solid arrow. Each improving flip that is induced by a relative importance preference statement (shown using dotted arrow) is annotated with the preference statement that induced the flip. The arrows are directed toward the preferred alternative in each flip. Note that $\delta(P^{TCP})$ contains improving flips corresponding to relative importance (p_2 in P^{TCP}), which is not present in $\delta(P^{CP})$.

In the case of the CI-net P^{CI}, the induced preference graph $\delta(P^{CI})$ includes flips induced by the (default) monotonicity rule (shown by solid arrows), and those induced by relative importance statements in the specification (shown by dotted arrows, annotated with the inducing preference statement).

Remark 2.18 We have so far seen how preference specifications in the languages of CP-nets, TCP-nets, CI-nets, and CP-theories can be expressed in the same syntactic framework, and the

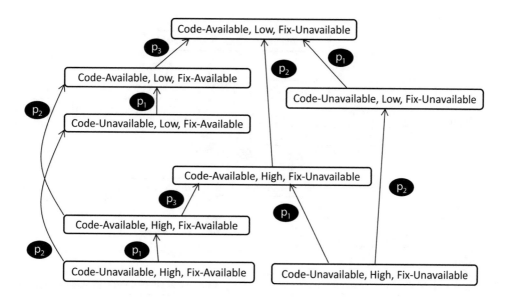

Figure 2.2: Induced preference graph $\delta(P^{CP})$ for the CP-net P^{CP}.

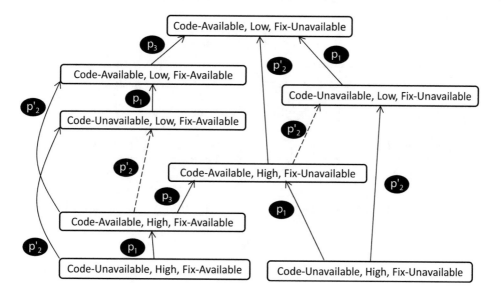

Figure 2.3: Induced preference graph $\delta(P^{TCP})$ for the CP-net P^{TCP}.

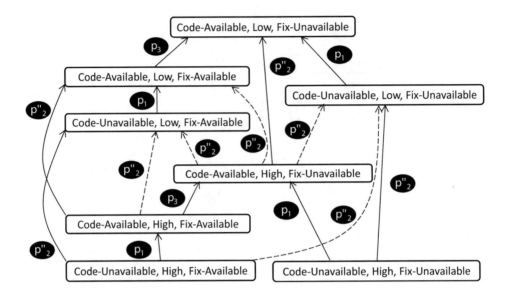

Figure 2.4: Induced preference graph $\delta(P^{CPT})$ for the CP-net P^{CPT}.

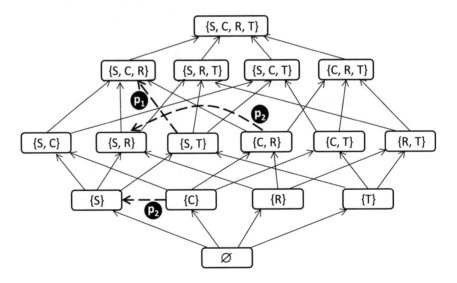

Figure 2.5: Induced preference graph $\delta(P^{CI})$ for CI-net P^{CI}.

ceteris paribus preference semantics of any preference specification in this framework. In the rest of the chapter, we consider only preference specifications in the introduced framework, rather than considering preference specification with respect to a particular language. The strategies we describe for preference reasoning and their implementation details will all be presented with respect to arbitrary preference specifications in this framework.

3.3   DOMINANCE AND CONSISTENCY IN QUALITATIVE PREFERENCE LANGUAGES

We now define the dominance of one alternative over another with respect to a preference specification P in terms of the improving flips induced by the preference statements in P.

Definition 2.19   Improving Flipping Sequence.   Given a preference specification P in any of the above languages and two alternatives α and β, a sequence of alternatives from β to α, namely $\beta = \gamma_1, \ldots \gamma_n = \alpha$, where $\gamma_1, \ldots \gamma_n$ are alternatives in $\mathcal{O}$, is said to be an *improving flipping sequence* [16–18, 94] with respect to (induced by) P from β to α if and only if for all $i, i + 1$ there exists an improving flip from γ_i to γ_{i+1} in the induced preference graph $\delta(P)$. We call n the *length* of the improving flipping sequence.

We observe that each path in the induced preference graph from one alternative to another corresponds to an improving flipping sequence. Moreover, there may be multiple distinct improving flipping sequences from one alternative to another.

Example 2.20   In Figure 2.2, there is only one improving flipping sequence of length 1 in $\delta(P^{CP})$ from $\beta = \langle$Code-Available, High, Fix-Unavailable$\rangle$ to $\alpha = \langle$Code-Available, Low, Fix-Unavailable$\rangle$. Figure 2.3 shows that there are two improving flipping sequences of lengths 1 and 2 from β to α in $\delta(P^{TCP})$ (including the one in $\delta(P^{CP})$); and Figure 2.4 shows that there are four improving flipping sequences of lengths 1, 2, 2, and 3 in $\delta(P^{CPT})$ from β to α (including those in $\delta(P^{CP})$ and $\delta(P^{TCP})$).

Definition 2.21   Dominance.   Given a preference specification P in any of the above languages and two alternatives α and β, we say that α *is preferred to* or *dominates* β, denoted $P \models \alpha \succ \beta$, if and only if there is an improving flipping sequence with respect to P from β to α.

Example 2.22   Dominance   Consider the TCP-net P^{CP}.

$\langle$Code-Available, Low, Fix-Unavailable$\rangle$**dominates**$\langle$Code-Unavailable, High, Fix-Available$\rangle$

because of the following improving flipping sequence:

$\langle$Code-Unavailable, High, Fix-Available$\rangle \rightarrow \langle$Code-Unavailable, Low, Fix-Available$\rangle$
$\rightarrow \langle$Code-Available, Low, Fix-Available$\rangle \rightarrow \langle$Code-Available, Low, Fix-Unavailable$\rangle$

Many applications require the user preferences to induce an acyclic preference graph over outcomes, as cycles indicate the notion of an outcome dominating (or being preferred to) itself. Such preferences are said to be *consistent*.

Definition 2.23   Consistency.   A preference specification P is said to be *consistent* if and only if there exists no improving flipping sequence (of length > 1) from any alternative to itself with respect to P.

A *consistent* preference specification P induces a *strict partial order* $\succ$ on the set $\mathcal{O}$ of alternatives, or equivalently $\delta(P)$ is acyclic whenever P is consistent.

Example 2.24   Consistency   According to the above definition, all the example preference specifications we have considered so far are consistent. Now consider changing the preference on A in the TCP-net P^{TCP} to include a conditional dependency on F. Let the resulting TCP-net be P_1^{TCP} as shown in Figure 2.6(a). The corresponding induced preference graph is shown in Figure 2.6(b). Observe that the improving flips induced by the preference on A from ⟨Code-Unavailable, High, Fix-Available⟩ and ⟨Code-Available, High, Fix-Available⟩ have been reversed with this change (reversed flips shown in bold), which induces cyclic preferences involving these alternatives. Hence, P_1^{TCP} is inconsistent.

In some settings, intentionally or unintentionally the stated preferences may induce cycles in the induced preference graph. We define three possible relationships between outcomes that are useful in such settings. Given two outcomes α and β: α may be preferred to β and not the other way around, or α may be preferred to β and vice versa, or α and β are not preferred to each other. Note that we are also considering relationships between outcomes resulting in inconsistencies in preferences.

Definition 2.25   Preference, Indifference, and Equivalence.   Given the preference specification P,

Preference. $\alpha \succ_P \beta$ if and only if there is an improving flipping sequence with respect to P from β to α but not from α to β. The existence of such a sequence corresponds to the presence of a path from β to α in the induced preference graph.

Indifference. α is indifferent to β and vice versa, denoted by $\alpha \not\bowtie_P \beta$, if and only if there is no improving flipping sequence with respect to P from α to β nor from β to α. In short, there is no path between α and β in $\delta(P)$.

Equivalence. α and β are equivalent, denoted by $\alpha \bowtie_P \beta$, if and only if either $\alpha = \beta$ or there are improving flipping sequences from α to β and from β to α. In short, α and β belong to the same strongly connected component (SCC) in $\delta(P)$.

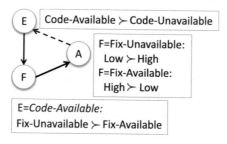

(a) TCP-net P_1^{TCP}

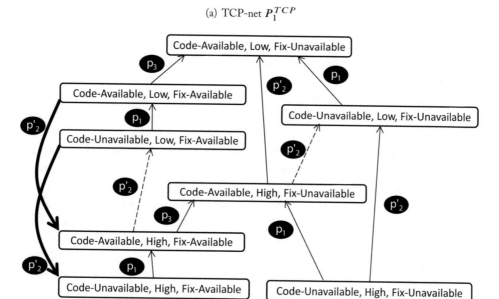

(b) Induced Preference Graph $\delta(P_1^{TCP})$

Figure 2.6: Inconsistent preference specification.

Example 2.26 For the inconsistent TCP-net's induced preference graph in Figure 2.6(b), ⟨Code-Available, Low, Fix-Unavailable⟩ is preferred to ⟨Code-Unavailable, High, Fix-Unavailable⟩; ⟨Code-Available, High, Fix-Available⟩ and ⟨Code-Unavailable, Low, Fix-Available⟩ are equivalent; and ⟨Code-Unavailable, High, Fix-Available⟩ and ⟨Code-Unavailable, High, Fix-Unavailable⟩ are indifferent according to the above definition. ⟨Code-Available, High, Fix-Available⟩ and ⟨Code-Unavailable, Low, Fix-Available⟩ are equivalent; and ⟨Code-Unavailable, High, Fix-Available⟩ and ⟨Code-Unavailable, High, Fix-Unavailable⟩ are indifferent according to the above definition.

4   COMPLEXITY OF REASONING

While it is useful to study qualitative preference languages due to their expressiveness, reasoning with expressive preferences is hard. This is because while preference statements in these languages are succinctly expressed (in terms of preferences among attributes and their valuations rather than directly among the exponentially many alternatives), their semantics is expressed in terms of reachability over the outcome space. In other words, the induced preference graph is exponential in the number of attributes and their domain sizes. In fact, Goldsmith et al. showed in [41] that consistency testing and dominance testing (even for consistent preferences) are PSPACE-complete for CP-nets, GCP-nets, and TCP-nets (even when the attributes have binary domains). The same is true of CI-nets [17] and CP-Theories [96]. Despite the fact that these languages are very expressive and succinct in terms of the representation scheme, the hardness of the key reasoning tasks of consistency checking and dominance testing limits the use of the above preference formalisms in practice.

Past work focused on identifying fragments of these languages, with corresponding restrictions on dependency structure of attributes (which restricts expressiveness), for which dominance can even be computed in polynomial time [16, 34]. In the next chapter, we first present a preference language that considers only unconditional preferences. Following that, we will turn our focus to making dominance testing practical when such restrictions are not imposed. We will focus on dominance testing as it is a basic building block[7] for preference reasoning.

Specifically, we examine algorithms and techniques for answering dominance and other preference reasoning queries using space and time that make their use feasible in practical applications. This requires the use of methods akin to what modern SAT solvers use to alleviate the difficulty of solving boolean satisfiability problems. The next chapter reviews model checking, a technique that, given a model of a system, exhaustively and automatically checks whether the model satisfies a given specification. In the following chapters we show how preference reasoning tasks can be formulated as model checking problems, allowing us to leverage the recent advances in model checking to develop practical approaches to preference reasoning.

---

[7]All the other relationships between outcomes can be derived from dominance; for example, strict preference is one way dominance, equivalence is two way dominance, and indifference corresponds to the case when neither holds.

CHAPTER 3

Model Checking and Computation Tree Logic

In Chapter 2, we discussed how preference queries for dominance and consistency can be answered using the induced preference graph that captures the semantics of different types of preference languages. We will present in Chapter 4 how the answers to these queries can be effectively computed and justified using the technique of model checking. In this chapter, we will present a brief overview of model checking for a specific type of temporal logic: *computation tree* or *branching tree temporal logic*. We will focus on the aspects of model checking that are necessary for the rest of the book—our presentation is, therefore, restricted to the context of this book.

- Computation-tree Temporal Logic (CTL): syntax and semantics

- Semantic model of CTL described as Kripke Structure

- Algorithm for checking whether Kripke structure is a semantic model of a given formula in the CTL logic

- A model checking engine, NuSMV.

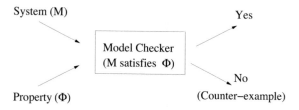

Figure 3.1: Process of model checking.

1 INTRODUCTION

Model Checking is an automated verification technique, which relies on mathematical logic, automata-theory, and graph algorithms to formally prove or disprove temporal properties of (hardware or software) systems. Broadly speaking, model checking requires two inputs: a finite graph, describing the dynamics of the system under consideration, and a formula in a suitable logic, *temporal logic*, describing the property to be verified. The semantics of the logical formula is described over states or paths in a graph, i.e., given a property in such a logic, the models for the property are graph structures. The objective of model checking is to find whether the graph describing the system dynamics is a model of the property and hence the name *model checking*. Two important aspects that make model checking attractive for verification purposes (besides providing formal guarantees of correctness) are (a) automation and (b) post-mortem analysis that brings out evidence as to how and why the provided input graph violated the property. Automation primarily stems from the fact that checking whether a finite graph representation of a system is a model for property can be implemented effectively using graph-exploration algorithms such as depth-first exploration and detection of strongly connected components [86]. The cause for violation, referred to as counterexample, can be automatically generated as well—when an exploration identifies the sub-graph that causes the graph being explored to not be a model of the given property (i.e., violate the property). Such counterexamples help to identify the errors in the system behavior or to understand whether or not the property being verified should be relaxed. Figure 3.1 illustrates this process.

The birth of model checking in its current form is attributed to the seminal work by Clarke and Emerson [28, 29] in 1981, and Queille and Sifakis [75] in 1982. However, work by many others has contributed significantly in developing the theory, the algorithms, and the implementation techniques used in model checking. For instance, the work by Pnueli on linear temporal logic [73], the results from Tarski-Knaster fixed point lemma [87], Kozen's work on propositional mu-calculus [56], and the contributions of many in developing efficient graph-exploration and automata-theoretic techniques [12, 90], to name a few, played significant roles in bringing model checking (and formal verification, in general) to the 21st century. While model checking has influenced primarily the area of verification of systems whose dynamics can be captured using finite graphs (even when the graphs have a very large number of states [22, 30, 60, 70]), a number of new techniques and algorithms have been also developed, investigated, and deployed for model checking systems (primarily software systems), whose behavior as a graph exhibits an infinite number of states (e.g., [1, 6, 8, 47–49, 74]).

In the following sections, we briefly describe the basics of model checking that are relevant in the context of this book.

2   KRIPKE STRUCTURE

Kripke structure [31] is a finite-state graph over entities and relations between entities, where the states denote the entities and the transitions/directed edges denote the entity-to-entity relationships. Such structures are often used to describe the semantics of modal logics where the modalities express the type of relationships among entities.

In the context of model checking, Kripke structure represents the dynamics of system behavior over discrete time—the states correspond to the configurations of the system and the tran-

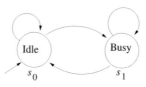

Figure 3.2: Kripke structure for printer.

sitions correspond to evolution of the system from one configuration to another. Formally, a Kripke structure description of a system is defined as follows:

Definition 3.1   Kripke Structure.   A Kripke structure $M = (S, S_0, T, L)$ is a finite-state graph where S is a finite set of states, $S_0 \subseteq S$ is a finite set of start states, $T \subseteq S \times S$ is a left-total transition relation such that $\forall s \in S : \exists s' \in S : (s, s') \in T$ and $L : S \rightarrow \mathcal{P}(AP)$ is a labeling function that maps each state $s \in S$ to a subset of propositions from a set of propositions AP.

We will denote $(s, s') \in T$ as $s \rightarrow s'$ for all $s, s' \in S$; we will call s the source state and s' the destination of the transition. Note that in Kripke structure every state in the Kripke structure has an outgoing transition. As a result, any path from any state in the Kripke structure can be of infinite length, where a path is described as a sequence of states such that each pair of successive states in the sequence is related by T. For any state $s \in S$, we will denote the set of all paths starting from s as $\texttt{Path}(s)$. We will use π and σ (with and without subscripts) to denote the a path, $\pi[0]$ to denote the states from where the path starts, and $\pi[i]$ to denote the i-th state in the path.

Example 3.2   Simple Printer Model   Consider the printer system described as a Kripke structure in Figure 3.2. It has two states: idle and busy. Initially, the printer is idle. The printer can remain idle (self-loop) or can move to busy state. At the busy state, the printer can remain busy or can move to idle state. The two states are denoted by s_0 and s_1, where s_0 is the start state. The label of the two states describe the "state" of the printer: $L(s_0) = \{\texttt{idle}\}$, $L(s_1) = \{\texttt{busy}\}$. Note that the Kripke structure of the printer is non-deterministic—the transitions from the states may non-deterministically move the system to a different state or to the state where it originated.

3   COMPUTATION TREE TEMPORAL LOGIC

Computation Tree Temporal Logic (CTL) [28] is a type of branching time logic which is used to express properties of the system dynamics described as Kripke structure, taking into consideration the branching behavior of the system; as noted before, the branching behavior comes into play due to possible non-determinism in the Kripke structure. These properties of interest can be viewed as the evolution of the system starting from some state of its Kripke structure description. In the following, we present a brief overview of the syntax and semantics of the CTL followed by the algorithms and complexity for verifying CTL properties against a given Kripke structure.

3.1   SYNTAX

The syntax of formula φ in CTL logic is described over propositional constant `true`, a set of propositions AP (in propositional logic), boolean operators (e.g., $\wedge, \vee, \neg$), and a set of temporal operators (EX, AX, EF, AF, EG, AG, EU, AU).

$$\varphi \rightarrow \text{true} \mid AP \mid \neg\varphi \mid \varphi \wedge \varphi \mid \varphi \vee \varphi \mid \text{A}\psi \mid \text{E}\psi$$

$$\psi \rightarrow \text{X}\varphi \mid \text{F}\varphi \mid \text{G}\varphi \mid \varphi \text{ U } \varphi$$

Intuitively, φ describes the properties over states in a Kripke structure and ψ denotes the properties over paths in the Kripke structure. The property `true` is satisfied in all states of a Kripke structure. The property over atomic propositions is satisfied only in the states which are labeled with the proposition. The negation of a CTL property is satisfied in states which do not satisfy the CTL property. The binary boolean operator $\wedge$ and $\vee$ have their natural semantics. The CTL property using the temporal operators can be read as follows:

A : **A**long all paths

E : **E**xists a path

X : in ne**X**t state

F : in **F**uture state

G : **G**lobally or in all states

U : **U**ntil

Note that, in a valid CTL property, the temporal operators are paired, where the first element in the pair is from {A, E} and the second element is from {X, F, G, U}. For instance, the property AX(p) is satisfied in a state s if and only if *along all paths* starting from s, *in the next state*, p is satisfied. In similar fashion, the property EF(p) is satisfied in a state s if and only if there *exists a path* from s, *in a future state*, p is satisfied.

3.2   SEMANTICS

As indicated above, the semantics of CTL property is described over a set of states in the Kripke structure that satisfy the property. We denote the semantics of the property φ in the context of a Kripke structure M as $[\![\varphi]\!]_M$. Formally, the semantics is described in Figure 3.3. For any $s \in [\![\varphi]\!]_M$,

$$[\![\text{true}]\!]_M = S$$

$$[\![p]\!]_M = \{s \mid s \in S \wedge p \in L(s)\}$$

$$[\![\neg\varphi]\!]_M = S - [\![\varphi]\!]_M$$

$$[\![\varphi_1 \wedge \varphi_2]\!]_M = [\![\varphi_1]\!]_M \cap [\![\varphi_2]\!]_M$$

$$[\![\varphi_1 \vee \varphi_2]\!]_M = [\![\varphi_1]\!]_M \cup [\![\varphi_2]\!]_M$$

$$[\![\text{EX}(\varphi)]\!]_M = \{s \mid \exists \pi \in \text{Path}(s) : \pi[1] \in [\![\varphi]\!]_M\}$$

$$[\![\text{AX}(\varphi)]\!]_M = \{s \mid \forall \pi \in \text{Path}(s) : \pi[1] \in [\![\varphi]\!]_M\}$$

$$[\![\text{EF}(\varphi)]\!]_M = \{s \mid \exists \pi \in \text{Path}(s) : \exists i \geq 0 : \pi[i] \in [\![\varphi]\!]_M\}$$

$$[\![\text{AF}(\varphi)]\!]_M = \{s \mid \forall \pi \in \text{Path}(s) : \exists i \geq 0 : \pi[i] \in [\![\varphi]\!]_M\}$$

$$[\![\text{EG}(\varphi)]\!]_M = \{s \mid \exists \pi \in \text{Path}(s) : \forall i \geq 0 : \pi[i] \in [\![\varphi]\!]_M\}$$

$$[\![\text{AG}(\varphi)]\!]_M = \{s \mid \forall \pi \in \text{Path}(s) : \forall i \geq 0 : \pi[i] \in [\![\varphi]\!]_M\}$$

$$[\![\text{E}(\varphi_1 \cup \varphi_2)]\!]_M = \{s \mid \exists \pi \in \text{Path}(s) : \exists i \geq 0 : \pi[i] \in [\![\varphi_2]\!]_M \wedge \forall j < i : \pi[j] \in [\![\varphi_1]\!]_M\}$$

$$[\![\text{A}(\varphi_1 \cup \varphi_2)]\!]_M = \{s \mid \forall \pi \in \text{Path}(s) : \exists i \geq 0 : \pi[i] \in [\![\varphi_2]\!]_M \wedge \forall j < i : \pi[j] \in [\![\varphi_1]\!]_M\}$$

Figure 3.3: Formal semantics of CTL with respect to Kripke structure $M = (S, S_0, T, L)$.

we will also write $M, s \models \varphi$ (in model M, the state s satisfies property φ). (We will omit M from the semantics-notation and satisfies-relation if the M is immediate from the context).

Example 3.3 Going back to Figure 3.2, the semantics of $[\![\text{EX}(\text{idle})]\!] = \{s_0, s_1\}$ as both the states has a path where in the next state the proposition idle is satisfied. On the other hand, $[\![\text{AX}(\text{idle})]\!] = \emptyset$.

Equivalences.   A number of equivalences between properties are immediate from the CTL semantics. This paves the way for identifying an *adequate set* of CTL operators—the minimal set of operators that is sufficient to express all possible CTL properties. For instance,

$$\neg\text{EX}(\varphi) \equiv \text{AX}(\neg\varphi)$$

The proof is as follows.

$$\llbracket \neg EX(\varphi) \rrbracket_M \;=\; S - \llbracket EX(\varphi) \rrbracket_M$$

$$=\; S - \{s \mid \exists \pi \in \text{Path}(s) : \pi[1] \in \llbracket \varphi \rrbracket_M \}$$

$$=\; \{s \mid \forall \pi \in \text{Path}(s) : \pi[1] \in S - \llbracket \varphi \rrbracket_M \}$$

$$=\; \{s \mid \forall \pi \in \text{Path}(s) : \pi[1] \in \llbracket \neg\varphi \rrbracket_M \}$$

$$=\; \llbracket AX(\neg\varphi) \rrbracket_M$$

The following equivalences hold as well. We leave the proofs as an exercise for the reader.

$$\neg EF(\varphi) \equiv AG(\neg\varphi) \quad \neg EG(\varphi) \equiv AF(\neg\varphi) \quad E(\text{true } U \; \varphi) \equiv EF(\varphi) \quad A(\text{true } U \; \varphi) \equiv AF(\varphi)$$

The equivalence of the negation of AU-property is more involved.

$$\llbracket \neg A(\varphi_1 \; U \; \varphi_2) \rrbracket_M \;=\; S - \{s \mid \forall \pi \in \text{Path}(s) : \exists i \geq 0 : \pi[i] \in \llbracket \varphi_2 \rrbracket_M \wedge \forall j < i : \pi[j] \in \llbracket \varphi_1 \rrbracket_M \}$$

$$=\; \{s \mid \exists \pi \in \text{Path}(s) : \forall i \geq 0 : \pi[i] \in \llbracket \neg\varphi_2 \rrbracket_M \vee \exists j < i : \pi[j] \in \llbracket \neg\varphi_1 \rrbracket_M \}$$

Therefore, a state s satisfies $\neg A(\varphi_1 \; U \; \varphi_2)$, i.e., s does not satisfy $A(\varphi_1 \; U \; \varphi_2)$ if and only if one of the following holds for some path from s:

1. φ_2 is satisfied in none of the states in the path

2. φ_1 is *not satisfied in all states before* the first state (in the path) that satisfies φ_2

The first item can be expressed in CTL as $EG(\neg\varphi_2)$. The second item can be expressed in CTL as $E(\neg\varphi_2 \; U \; (\neg\varphi_1 \wedge \neg\varphi_2))$. Therefore,

$$\neg A(\varphi_1 \; U \; \varphi_2) \equiv EG(\neg\varphi_2) \vee E(\neg\varphi_2 \; U \; (\neg\varphi_1 \wedge \neg\varphi_2))$$

Using the same logic, the negation of $E(\varphi_1 \; U \; \varphi_2)$ demands that along all paths from a state condition 1 or condition 2 is satisfied. Note that this implies in some paths condition 1 may be satisfied and in some other condition 2 is satisfied (i.e., it is not necessary that the same condition is satisfied in all the paths). Such a disjunctive condition over all paths from a state cannot be described in CTL. As a result, the negation of EU-properties is not expressible in CTL using operators other than EU.[1]

Given the above equivalences, one of the adequate set of CTL operators is $\{\neg, \vee, EX, EG, EU\}$. The temporal properties using

- AX can be expressed using $\neg$ and EX

- AF can be expressed using $\neg$ and EG

- AG can be expressed using $\neg$ and EF

- EF can be expressed using EU

---

[1]Formally proving the inadequacy of expressing negation of EU-properties using other CTL operators is beyond the scope of this book. For details, the reader is referred to [5, 50].

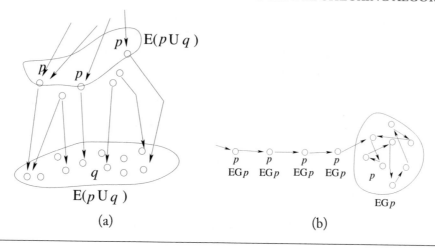

Figure 3.4: Illustration for the computation of the semantics for (a) E(p U q) and (b) EG(p).

4   MODEL CHECKING ALGORITHM

As noted before, model checking amounts to automatically verifying where the dynamic behavior of a system conforms to a given (desired) property. More specifically, in model checking the system behavior is described using Kripke structure and the properties of interest are expressed in the temporal logic, such as CTL. In CTL model checking, the objective is to automatically verify whether all start states of the Kripke structure (describing some system's behavior) belongs to the semantics of a given CTL property.

We consider the adequate set of CTL operators (as presented above) and present a brief overview of the model checking algorithm for computing the semantics of CTL properties described using the adequate set of operators. The steps of the algorithm are as follows. Given a Kripke structure M and the CTL property φ:

1. if $\varphi = \mathtt{true}$, then the set of all states in M is returned as the result.

2. if $\varphi = p$ (atomic proposition), then set of all states s in M where $p \in L(s)$ is returned as the result.

3. if $\varphi = \neg\varphi'$, then the algorithm recursively computes the semantics of φ'. Any state that is not present in the semantics of φ' is returned as a result.

4. if $\varphi = \varphi_1 \vee \varphi_2$, then the algorithm recursively computes the semantics of both φ_1 and φ_2. The set of states obtained from the union of those semantics is returned as the result.

5. if $\varphi = \mathtt{EX}(\varphi')$, then the algorithm computes the semantics of φ' and the set of all states, which can reach a state in the semantics of φ' in one step, is returned as the result.

6. if $\varphi = \text{E}(\varphi_1 \ \text{U} \ \varphi_2)$, then the algorithm first computes the semantics of φ_2 and φ_1. The set containing the states in the semantics of φ_2 and the ones that are in the semantics of φ_1 and can reach states in the semantics of φ_2 is returned as the result. This computation involves backward depth-first exploration of M from the states in the semantics of φ_2.

7. if $\varphi = \text{EG}(\varphi')$, then the algorithm first computes the semantics of φ' and removes all states from M that are not the semantics. In the second step, strongly connected components over the remaining states (say, S') are computed. The set of all states among the S' that can reach these strongly connected components are returned as the result.

Figure 3.4 presents the backward exploration of Kripke structure for computing the semantics of EU and EG-properties. The enclosed set of states at the bottom of Figure 3.4(a) satisfies the proposition q, and therefore, satisfies the property $\text{E}(p \ \text{U} \ q)$. Backward exploration from this set of states proceeds through the states that satisfy the proposition p. In case of $\text{EG}(p)$ (Figure 3.4(b)), the exploration starts from the strongly connected component containing only the states that satisfy p.

The time complexity of the model checking algorithm is $O(|\varphi|.(|S| + |T|))$, where $|\varphi|$ denotes the size of the CTL property (controlling the number of recursive steps in the algorithm), and $|S|$ and $|T|$ are the total number of states and transitions, respectively, in M. Each of the algorithms can be explained as follows. The complexity of algorithm steps 1–4 is of the order of $|S|$. The complexity of step 5 is $|T|$ as it is necessary to explore all possible outgoing transitions from every state in M. The complexity for each step 6 and 7 is of the order of $|S| + |T|$ owing to the complexity of depth-first exploration and that of strongly connected component computation [86].

We refer the readers to [50] and [5] for the details of CTL model checking algorithm.

5 NUSMV MODEL CHECKER

NuSMV [65], a variant of SMV [85], takes as input system behavior described as guarded statements (enabled only when certain conditions expressed as a boolean formula are satisfied), temporal properties expressed in the logic of CTL, and automatically translates the system behavior into Kripke structure and model checks whether the start states of the structure satisfy the given properties.

While the model checker relies on the algorithm described above for computing semantics of CTL properties, it uses a special type of data structure to represent sets of states and transitions in a compact, succinct, and canonical fashion. The data structure is referred to as the Binary Decision Diagram (BDD) [2, 21], which has been efficiently and effectively used to represent propositional logic formulas. NuSMV (and SMV) utilizes this technique to represent sets of states and transitions expressed as propositional logic formulas. All computations for model checking are performed at the level of sets rather than explicitly at the level of entities (states or transitions) in a state. Such computations are referred to as *symbolic* or set-based and the model checker is called "symbolic model verifier."

The technique for symbolic verification was introduced in [22] and [60] to address specifically the problem of verifying Kripke structure with very large state-space. Note that given a set of propositions AP, the number of states in the Kripke structure can be potentially $2^{|AP|}$, i.e., the state-space could grow exponentially with respect to the number of the propositions. This makes the algorithm for model checking prohibitively space inefficient if every state is explicitly saved. As labeling function maps each state to a set of propositions, a state can be described as a conjunction of propositions that hold in that state, and the set of states can be described as a disjunction of conjunctive formulas. It turns out that such logical representation of sets can be compact and loss-less.

Consider for instance, a Kripke structure containing four states such that $L(s_0) = \{p\}, L(s_1) = \{q\}, L(s_2) = \{p, q\}, L(s_3) = \emptyset$. The set of all states can be represented by the propositional constant true. The set $\{s_0, s_1, s_2\}$ is represented by $p \vee q$.

A Binary Decision Diagram precisely represents propositional logic formulas. A BDD typically has one root and two leaf-nodes. The leaf-nodes are true and false nodes. Each intermediate node (including the root) in the BDD is a propositional variable and its outgoing edges denote its values: true or false. A path in the BDD from the root to the true leaf-node captures an assignment to the propositional variables for which the formula evaluates to true.

There are two important features of BDD. First, it does not contain any *redundant node*. A node is redundant if the valuation of the formula does not depend on the valuation of a variable at that node for some truth-assignment of the other variables. Second, the BDD does not contain any *duplicate nodes*. A node is duplicate of another (each corresponding to the same variable) if their true and false edges lead to the same node in the decision diagram. The BDD with these features is also referred to as *Reduced Binary Decision Diagram*.

Example 3.4   Consider the Kripke structure with four states: s_0, s_1, s_2, s_3. The transitions are $s_0 \to s_1, s_1 \to s_2, s_2 \to s_3, s_3 \to s_2$. The labeling function maps the states to sets of propositions that are satisfied in the states: $L(s_0) = \{p\}, L(s_1) = \{q\}, L(s_2) = \{p, q\}, L(s_3) = \emptyset$. The propositional logic representation of this Kripke structure is as follows:

$$[(p \wedge \neg q \wedge \neg p' \wedge q') \vee (\neg p \wedge q \wedge p' \wedge q') \vee (p \wedge q \wedge \neg p' \wedge \neg q') \vee (\neg p \wedge \neg q \wedge \neg p' \wedge \neg q')$$

Each disjunct in the above formula describes a transition in the Kripke structure. For each transition, the propositions that are satisfied in the source-state belong to the set $\{p, q\}$, while the propositions that are satisfied at the destination-state are denoted by their primed-version. For instance, $(p \wedge \neg q \wedge \neg p' \wedge q')$ denote the transition from a state where only p is satisfied to a state where only q is satisfied.

Figure 3.5 illustrates the BDD representation of the transition relation. The dotted edges from a node represent the case when the valuation of the variable at the node is considered to be *false*; the solid edges correspond to the case when the variable is considered to be *true*. The leaf

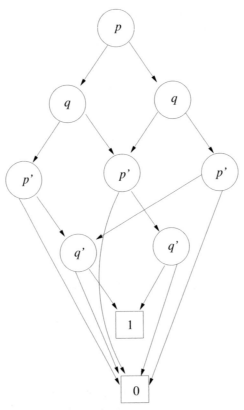

Figure 3.5: BDD representation of transition relation for Example 3.4.

nodes indicate the value of the propositional logic formula for each assignment of the variables (starting from the variable at the root node). For instance, if p and p' are both true, then the propositional logic formula evaluates to false.

In Figure 3.5, we used the ordering $[p, q, p', q']$. We leave it to the reader to draw the BDD for the same propositional logic formula with the variable ordering $[p, p', q, q']$.

The primary challenge in dealing with BDDs is in finding the order in which the variables should be tested to obtain the most compact BDD representation of a propositional formula. The problem, in general, is NP-complete [10]. A number of heuristics have been developed which rely on domain-knowledge (or structure) of the propositional formula and/or on learning dynamically the ordering of variables that is likely to result in smaller BDD [42].

Example 3.5   Consider the reduced ordered BDDs over the variables p, q, r and their primed versions in the Figure 3.6. For clarity of the figures, only the edges leading to the "true"-valuation (1 leaf-node) of the formula represented by the BDDs are presented.

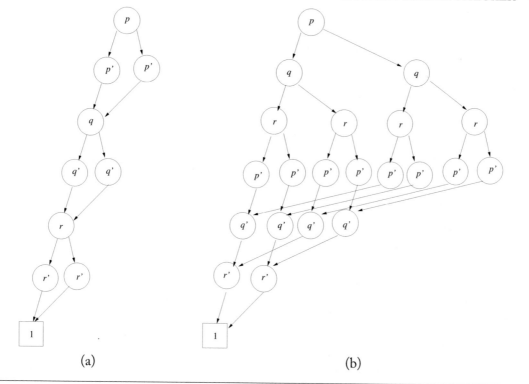

Figure 3.6: Impact of variable ordering on BDD size.

Careful observation will reveal that these two BDDs are representing the same propositional formula, which is satisfiable only when the valuation of the variable and their primed version are equal (in terms of transition relation in a Kripke structure, this is a self-loop). The ordering used in Figure 3.6(a) results in a BDD that is significantly smaller (and less complex) compared to the one resulting from the ordering used in Figure 3.6(b).

For details on Binary Decision Diagram and Symbolic Model Checking, please see Logic in Computer Science [5, 50].

5.1   NUSMV LANGUAGE & COUNTEREXAMPLES

We refer the reader to review the NuSMV manual available at http://nusmv.fbk.eu/. Here we present the basic features and constructs of the language that are necessary for explaining the modeling strategy used in the subsequent chapters.

A NuSMV specification of a system model consists of modules identified using the keyword MODULE. Each module describes a process in the system and can be composed synchronously or asynchronously. The top-level module is referred to as the main module, and every NuSMV spec-

```
MODULE main

VAR
    st: {0, 1};      -- 0: idle, 1: busy


ASSIGN
    init(st) := 0;  -- initial state is 0

    next(st) := case
          st = 0: {0, 1}; -- if st is 0, then next value of st can be 0 or 1
                st = 1: {0, 1};
          TRUE : st;     -- if no other guard holds, the st value remains unchanged
                esac;

SPEC---CTL specification
    AX (idle)
```

Figure 3.7: Printer model in NuSMV.

ification must contain this module. Each module has several building blocks—each responsible for describing certain features of the process represented by the module.

The VAR block declares the variables present in the module. In NuSMV, all variables have finite-domain (e.g., enumerated type variables or boolean variables).

The ASSIGN block deals with initialization of the declared variables and with specification of how the variables can be updated for each transition in the module. The specification of variable updates are described using multiple guarded-assignments. For instance, in Figure 3.7, the variable st is initialized to 0. The update specifications for this variable are described using three guarded-assignments. If the value of the variable st in the current state is 0, then as a result of a transition, the value of st in the destination state can be either 0 or 1.

It is not necessary to initialize all declared variables or to specify the updates to all declared variables. For uninitialized variables, the model checker considers all possible initializations (as per the variable domain) and similarly, for variables whose updates are not specified, the model checker considers all possible value-updates.

For describing the CTL properties, NuSMV has a block SPEC. The syntax of CTL is the same as the ones described in the previous sections. The model checker NuSMV considers all the CTL properties described under SPEC and verifies whether each one of them is satisfied by all the start states of the specified model. If a property is not satisfied, the model checker generates a counterexample illustrating the violation of the property.

Note that for properties of the form $AG(\varphi)$, violation results in a counterexample described as a sequence of states in the system-model ending in a state that satisfies $\neg\varphi$. This is because the counterexample must satisfy $EF(\neg\varphi)$ (recall the equivalences between CTL properties). For properties of the form $AF(\varphi)$, violation results in a counterexample described as a sequence of states in the system-model, where each state satisfies $\neg\varphi$ and the last state in the sequence has a transition back to some state in the sequence. This is because the counterexample must satisfy $EG(\neg\varphi)$. On the other hand, for the violation of properties $EF(\varphi)$ and $EG(\varphi)$ NuSMV simply shows the start state that is responsible for the violation. This is because the counterexample for both these types of properties are trees rooted at some start state (NuSMV developers have decided to present one start state).

CHAPTER 4

Dominance Testing via Model Checking

We have discussed in Chapter 2 several preference languages where the basic building blocks are conditional intra-variable preferences or CP-nets. We also saw that the complexity of dominance testing, a fundamental preference reasoning task, is in general intractable. To cope with this hardness, in this chapter we discuss how model checking can be used to compute dominance, and how this approach is applicable to a broad class of conditional preference languages including CP-nets, TCP-nets, CP-theories, and CI-nets.

Before we describe the model checking approach, a natural question arises—is there a restricted preference language of unconditional preferences that is tractable and can be used in place of CP-nets or TCP-nets? To address this question first, Section 1 describes a preference language [82] that restricts the language's expressivity to unconditional preferences. While the model checking approach is general and can also be used to reason about unconditional preference languages, we will see that there is merit in studying the unconditional case separately because dominance testing can be done in polynomial time for the restricted case. Sections 2 and 3 then describe the details of the model checking approach to dominance testing.

1 DOMINANCE TESTING OF UNCONDITIONAL PREFERENCES

We first focus on unconditional preferences as a building block for expressing more complicated preferences. We describe a simple preference language [82] that allows both intra-variable importance preferences and relative importance preferences without conditions. We will see that dominance testing in the language presented in the beginning of this chapter will be tractable in polynomial time, in contrast to the languages considered so far, wherein it is intractable.

One can imagine several applications where preferences are unconditional. One example is sustainable design [83], where a user may be interested in assessing the sustainability of alternative designs of buildings, pavements, etc., that are assembled from multiple components. Multi-attribute selection problems in engineering applications such as the selection of the best materials for a mechanical design [3] or the selection of the best design for the construction of pavements [79] are also candidates for application of the approach presented in this chapter.

We consider a qualitative preference language $\mathcal{L}$ for specifying (over a set $\mathcal{X} = \{x_1 \ldots x_n\}$ of variables with domain D_i for each x_i): (a) unconditional intra-variable preferences $\succ_i$ that are strict partial orders (i.e., irreflexive and transitive relations) over D_i; and (b) unconditional relative importance preferences that are strict partial orders over $\mathcal{X}$ that can be used to express and reason about preferences in settings such as the above.

1.1   SYNTAX OF $\mathcal{L}$

Following the notation we introduced earlier, $\mathcal{L}$ includes unconditional preference statements of the form $x \succ_i x'[\mathcal{Z}]$ such that $x, x' \in D_i$, and $\{x_i\} \notin \mathcal{Z}$. Here, the set $\mathcal{Z}$ of variables are relatively less important than x_i, i.e., $x_i \rhd x_j$ for each $x_j \in \mathcal{Z}$. However, the language $\mathcal{L}$ does not include statements specifying conditional relative importance. Additionally, because $\rhd$ is assumed to be a binary (strict partial order) relation, $|\mathcal{Z}| = 0$ or 1.

We next consider several alternative semantics for the unconditional preference language $\mathcal{L}$ in terms of a binary preference relation $\succ$ (dominance) over alternatives, which is derived from the input preferences $\{\succ_i\}$ and $\rhd$.

1.2   SEMANTICS OF $\mathcal{L}$

We now specify the semantics for dominance in the language $\mathcal{L}$. For the unconditional intra-variable preferences ($\succ_i$ for each x_i) and relative importance preferences ($\rhd$) specified in a preference specification of $\mathcal{L}$, we define dominance of α over β in terms of a first order logic formula parameterized by the alternatives α and β, and preferences $\succ_i$ and $\rhd$ such that the satisfiability of the formula determines whether or not α dominates β. We denote by $\succ^\bullet$ the dominance relation induced by the satisfiability of the formula over alternatives.

We first define a preference relation $\succeq_i$ (for each variable x_i) that is derived from $\succ_i$, which is necessary to give the semantics of $\mathcal{L}$.

Definition 4.1 $\succeq_i$. $\forall u, v \in D_i : u \succeq_i v \Leftrightarrow u = v \vee u \succ_i v$

Note that since $\succ_i$ is a strict partial order, i.e., irreflexive and transitive, it easy to show that $\succeq_i$ is reflexive and transitive.

We next define the dominance between any pair of alternatives using a logic formula, for unconditional intra-variable ($\succ_i$, $\succeq_i$) and relative importance ($\rhd$) preferences.

Definition 4.2 Dominance in $\mathcal{L}$.   Given a preference specification in $\mathcal{L}$ defining intra-variable preferences $\{\succ_i\}$ and relative importance preference $\rhd$, and a pair of alternatives α and β, we say that α **dominates** β, denoted $\alpha \succ^\bullet \beta$ whenever the following holds.

$$\alpha \succ^\bullet \beta \Leftrightarrow \quad \exists x_i : \alpha(x_i) \succ_i \beta(x_i) \wedge$$
$$\forall x_k : (x_k \rhd x_i \vee x_k \sim_\rhd x_i)$$
$$\Rightarrow \alpha(x_k) \succeq_k \beta(x_k)$$

x_i is called the witness of the relation.

Intuitively, this definition of dominance of α over β (i.e., $\alpha \succ^{\bullet} \beta$) requires that with respect to at least one attribute, namely the witness, α is *preferred* to β. Further, it requires that for all attributes that are relatively more important or indifferent (incomparable) with respect to relative importance to the witness, α *either equals or is preferred to* β. In Example 4.11, $\alpha \succ^{\bullet} \beta$, with x_1 serving as the witness.

1.3   PROPERTIES OF UNCONDITIONAL DOMINANCE RELATION

We now proceed to analyze some properties of $\succ^{\bullet}$. Specifically, we would like to ensure that $\succ^{\bullet}$ has two desirable properties of preference relations: irreflexivity and transitivity, which make it a strict partial order. First, it is easy to see that $\succ^{\bullet}$ is irreflexive, due to the irreflexivity of $\succ_i$ (since it is a partial order).

Proposition 4.3   Irreflexivity of $\succ^{\bullet}$ $\forall \alpha : \alpha \not\succ^{\bullet} \alpha$.

The above proposition ensures that the dominance relation $\succ^{\bullet}$ is strict over compositions. In other words, no composition is preferred over itself. Regarding transitivity, we observe that $\succ^{\bullet}$ is not transitive when $\succ_i$ and $\rhd$ are both arbitrary strict partial orders, as illustrated by the following example.

Example 4.4   Let $\mathcal{X} = \{x_1, x_2, x_3, x_4\}$, and for each $x_i \in \mathcal{X} : D_i = \{a_i, b_i\}$ with $a_i \succ_i b_i$. Suppose that $x_1 \rhd x_3$ and $x_2 \rhd x_4$. Let $\alpha = \langle a_1, a_2, b_3, b_4 \rangle$, $\beta = \langle b_1, a_2, a_3, b_4 \rangle$ and $\gamma = \langle b_1, b_2, a_3, a_4 \rangle$. Clearly, we have $\alpha \succ^{\bullet} \beta$ (with x_1 as witness), $\beta \succ^{\bullet} \gamma$ (with x_2 as witness), but there is no witness for $\alpha \succ^{\bullet} \gamma$, i.e., $\alpha \not\succ^{\bullet} \gamma$ according to Definition 4.2.

Because transitivity of preference is a necessary condition for rational choice [39, 62], we proceed to investigate the possibility of obtaining such a dominance relation by restricting $\rhd$. In particular, we find that $\succ^{\bullet}$ is transitive when $\rhd$ is restricted to a special family of strict partial orders, namely *interval orders* as defined below.

Definition 4.5   Interval Order.   A binary relation $\mathbf{R} \subseteq \mathcal{X} \times \mathcal{X}$ is an interval order iff it is irreflexive and satisfies the *ferrers* axiom [37]: for all $x_i, x_j, x_k, x_l \in \mathcal{X}$, we have:
$$(x_i \, \mathbf{R} \, x_j \wedge x_k \, \mathbf{R} \, x_l) \Rightarrow (x_i \, \mathbf{R} \, x_l \vee x_k \, \mathbf{R} \, x_j)$$

Restricting $\succ^{\bullet}$ to an interval order is necessary and sufficient for the transitivity of $\succ^{\bullet}$ as stated below, which was proved in [84].

Theorem 4.6 $\succ^{\bullet}$ *is a strict partial order when intra-attribute preferences $\succ_i$ are arbitrary strict partial orders and relative importance $\rhd$ is an interval order.*

The above theorem applies to all partially ordered intra-variable preferences and a wide range of relative importance preferences including total orders, weak orders, and semi orders [37] which are all interval orders. Having seen in Example 4.4 that the transitivity of $\succ^{\bullet}$ does not

Figure 4.1: A $2 \oplus 2$ substructure, not an interval order.

necessarily hold when $\rhd$ is an arbitrary partial order, a natural question that arises here is whether there is a condition *weaker* than the interval order restriction on $\rhd$ that still makes $\succ^\bullet$ transitive. The answer turns out to be negative, which we show next. We make use of a characterization of interval orders by Fishburn in [37], which states that $\rhd$ is an interval order if and only if $2 \oplus 2 \not\subseteq \rhd$, where $2 \oplus 2$ is a relational structure shown in Figure 4.1. In other words, $\rhd$ is an interval order if and only if it has *no subrelation* whose structure is isomorphic to the partial order structure shown in Figure 4.1.

Theorem 4.7 *For arbitrary partially ordered intra-attribute preferences $\succ^\bullet$ is transitive only if relative importance $\rhd$ is an interval order.*

Proof. Assume that $\rhd$ is not an interval order. This is true if and only if $2 \oplus 2 \subseteq \rhd$. However, we showed in Example 4.4 that in such a case $\succ^\bullet$ is not transitive. Hence, $\succ^\bullet$ is transitive only if relative importance $\rhd$ is an interval order. $\square$

The preceding theorem holds for a wide range of relative importance preferences, including total orders, weak orders, and semi orders [37] which are all interval orders. Hence the language $\mathcal{L}$ can be used for decision making in these settings.

1.4 COMPLEXITY OF DOMINANCE TESTING IN $\mathcal{L}$

Dominance testing in $\mathcal{L}$ amounts to evaluating the satisfiability of the quantified boolean formula corresponding to $\alpha \succ^\bullet \beta$ (see Definition 4.2). This can be done in $O\big(m^2(m^4 + n^4)\big)$ time, where $m = |\mathcal{X}|$ is number of variables and $n = max_{x_i \in \mathcal{X}}|D_i|$ is size of the domains of variables. This is clearly polynomial time, and hence it makes $\mathcal{L}$ a useful alternative in settings where runtime complexity is important, and where the preferences can be stated unconditionally.

1.5 EXPRESSIVENESS

We now compare the expressiveness of $\mathcal{L}$ to that of some CP-languages we have introduced earlier. We use $\mathcal{L}_?$ where ? corresponds to the language we want to compare $\mathcal{L}$ against, namely CP (CP-nets), TCP (TCP-nets), $CPTh$ (CP-theory), CI (CI-net). We make the following observations:

- $\mathcal{L}$ is neither more expressive nor less expressive compared to $\mathcal{L}_{CP}$. $\mathcal{L}$ allows the expression of relative importance while $\mathcal{L}_{CP}$ does not; and $\mathcal{L}_{CP}$ allows the expression of conditional intra-variable preferences while $\mathcal{L}$ does not.

- $\mathcal{L}$ is less expressive than $\mathcal{L}_{TCP}$ because it does not allow the expression of conditional intra-variable preferences and relative importance.

- When $\mathcal{L}_{TCP}$ is restricted to unconditional intra-variable and unconditional relative importance preferences, its expressiveness is the same as that of $\mathcal{L}$.

- $\mathcal{L}_{CPTh}$ is more expressive than $\mathcal{L}_{CP}$ and $\mathcal{L}_{TCP}$ [93, 94], and therefore is more expressive than $\mathcal{L}$ as well.

- $\mathcal{L}_{CI}$ is neither more nor less expressive than $\mathcal{L}$ because the former can express relative importance among sets of attributes (which the latter cannot) and the latter can express any arbitrary intra-attribute preference relation over its variables (which the former cannot).

Comparing $\mathcal{L}$ with CP-Theories

We will first investigate the relationship between $\mathcal{L}$ and CP-theories, as some information about the relationship between $\mathcal{L}$ and TCP-nets can be inferred from this. Santhanam et al. [82] showed that the semantics of $\mathcal{L}$ is subsumed by the semantics of the CP-theories formalism.

Theorem 4.8 *The preference semantics of $\mathcal{L}$ is subsumed by the preference semantics of CP-theories, i.e., $\succ^{\bullet} \subseteq \succ^{\blacksquare}$.*

The above establishes that $\succ^{\bullet}$ is included in $\succ^{\blacksquare}$. To determine whether the other side of inclusion holds, we consider the following example.

Example 4.9 Consider again the example preferences from Example 4.4, where $\alpha = \langle a_1, a_2, b_3, b_4 \rangle$, $\beta = \langle b_1, a_2, a_3, b_4 \rangle$, and $\gamma = \langle b_1, b_2, a_3, a_4 \rangle$ with $\alpha \succ^{\bullet} \beta$ (with x_1 as witness), $\beta \succ^{\bullet} \gamma$ (with x_2 as witness), but $\alpha \not\succ^{\bullet} \gamma$ according to Definition 4.2. However, there exists a sequence of improving flips from γ to α, namely γ, β, α; hence $\alpha \succ^{\blacksquare} \gamma$.

The above example shows that $\succ^{\blacksquare} \subseteq \succ^{\bullet}$ does not hold in general. However, observe that $\succ^{\bullet}$ holds for each consecutive pair of alternatives in the swapping sequence. Hence, since $\succ^{\bullet}$ is transitive, it must be possible to show that $\succ^{\blacksquare} \subseteq \succ^{\bullet}$. The following result was proved by Santhanam et al. [82] using Theorem 4.6, which relates the interval order property of $\rhd$ to the transitivity of $\succ^{\bullet}$.

Theorem 4.10 *The preference semantics of CP-theories are subsumed by the preference semantics of $\mathcal{L}$, i.e., $\succ^{\blacksquare} \subseteq \succ^{\bullet}$, whenever the relative importance preference relation $\rhd$ is an interval order.*

Comparing $\mathcal{L}$ with TCP-nets

We now investigate the relationship between $\succ^{\circ}$ and $\succ^{\bullet}$. In Example 4.4, γ, β, α forms an improving flipping sequence from γ to α, resulting in $\alpha \succ^{\circ} \gamma$. However, $\alpha \not\succ^{\bullet} \gamma$. Since $\succ^{\bullet}$ holds for each pair of consecutive alternatives in a flipping sequence supporting a dominance $\alpha \succ^{\circ} \beta$, we have $\succ^\circ \subseteq \succ^\bullet$ when $\succ^\bullet$ is transitive. The other side of the inclusion is negated by Example 4.11, where $\alpha \succ^\bullet \beta$ but $\alpha \not\succ^\circ \beta$.

Example 4.11   Let $\mathcal{X} = \{X, Y, Z\}$ and $D_X = \{x_1, x_2\}$; $D_Y = \{y_1, y_2\}$; $D_Z = \{z_1, z_2\}$. Suppose that the intra-variable preferences are given by $x_1 \succ_X x_2$, $y_1 \succ_Y y_2$, and $z_1 \succ_Z z_2$, and the relative importance among the variables is given by $X \rhd Y$ and $X \rhd Z$. Given two alternatives $\alpha = \langle x_1, y_2, z_2 \rangle$ and $\beta = \langle x_2, y_1, z_1 \rangle$, there is **no** improving flipping sequence from α to β or vice versa with respect to TCP-net semantics. Therefore, $\alpha \not\succ^\circ \beta$ and $\beta \not\succ^\circ \alpha$.

This shows that neither of the preference semantics of TCP-nets or $\mathcal{L}$ subsumes the other in general; however when relative importance is an interval order, the preference semantics of TCP-nets subsumes the preference semantics of $\mathcal{L}$.

We so far described preference language $\mathcal{L}$ where all intra-variable and relative importance preferences are unconditional, whose semantics for dominance is in the form of the satisfaction of a quantified boolean formula (QBF). The complexity of checking whether a QBF formula is satisfiable can be done in polynomial time, which makes $\mathcal{L}$ a useful and tractable approximation for the preference semantics of TCP-nets and CP-theories earlier discussed according to *ceteris paribus* semantics introduced in Chapter 2. $\mathcal{L}$ is useful in several engineering applications such as materials design [3], sustainable building design [83], and pavement engineering [79].

2   PREFERENCE REASONING VIA MODEL CHECKING

We now turn to the more general case of conditional preference languages. Recall from Chapter 2 (see Definition 2.21) that an alternative α dominates another alternative β with respect to preference specification P if and only if the node corresponding to α in the preference graph $\delta(P)$ induced by P is reachable from the node corresponding to β. That is, dominance can be decided by performing reachability analysis of $\delta(P)$. If the preference specification P is consistent, then the corresponding $\delta(P)$ is guaranteed to be acyclic because it represents an irreflexive and transitive dominance relation on the set of outcomes. For the rest of the discussion in this chapter, we will assume P to be consistent to make things simple to explain and understand, although the same technique presented here will also work for dominance testing with inconsistent preferences.

As dominance can be decided via reachability analysis over (induced preference) graph, it becomes a perfect candidate application for model checking tools such as NuSMV (see Chapter 3). First, the interpretation of preference statements in the *ceteris paribus* languages or preference semantics in the form of induced preference graph can be neatly represented using *Kripke structures* [56] (for details see Chapter 3). Secondly, this encoding can be done without having to construct the induced preference graph explicitly, because the set of (possibly exponential) flips induced by a preference statement can be succinctly and implicitly encoded in terms of change of values of (state-)variables in the Kripke structure. Thirdly, as we will see, preference queries like dominance testing, which can be cast as reachability testing on the induced preference graph, can be easily translated into verifiable properties in a temporal logic language. Fourthly, we can take advantage of existing algorithms and tools for verifying temporal properties of models to answer preference queries like dominance. Finally, the model checker provides counterexamples to temporal properties that are not verified, which gives us evidence from the preference semantics to justify the answer. In this chapter, we will see all of the above aspects of model checking used in preference reasoning.

The model checking approach for dominance testing we are about to describe involves addressing two questions:

1. How to encode the preference statements as an input specification of a system in a model checker (we use NuSMV [26]) such that dynamics/evolution of the system (i.e., the Kripke structure) explored by the model checker corresponds to the preference graph ($\delta(P)$) induced by the preference statements?

2. How to express a query regarding the dominance of an alternative (α) with respect to another (β) in the form of a test of reachability of α from β in the Kripke structure corresponding to the induced preference graph?

We encode the preference specification P as the transition relations of a system specification to generate the Kripke model K_P corresponding to $\delta(P)$ as follows. The preference variables of the preference specification P are mapped to the state variables of the K_P in a model checker. The improving flips (see Definition 2.21, Chapter 2) are directly encoded as transition relations in K_P (in the language of the model checker). This ensures that the state space of the model K_P explored by the model checker corresponds to $\delta(P)$.

Dominance queries over P are then encoded as temporal logic properties in Computation Tree temporal logic (CTL; see Section 3 in Chapter 3) over the state space of the model K_P. This allows us to take advantage of all the specialized data structures (e.g., Binary Decision Diagrams [21]) and algorithms available in the model checking engine to efficiently verify the satisfiability of the temporal logic properties (over temporal operator such as EF) on K_P and as a result verify dominance of outcomes in the corresponding induced preference graph $\delta(P)$.

2.1   KRIPKE STRUCTURE ENCODING OF INDUCED PREFERENCE GRAPH

Given a preference specification P, the induced preference graph $\delta(P)$ is first encoded as a Kripke structure $K(P)$ (see Section 2 in Chapter 3) which is the standard input model understood by symbolic model checkers such as NuSMV and Cadence SMV.

The state space of $K(P)$ is specified using the valuations of the set of preference variables, namely X; we will often refer to them as *state* variables of the Kripke structure as each state $s \in S$ in $K(P)$ is defined by a unique assignment to X. We will use sub/super-scripted versions of s to represent specific states in $K(P)$ and denote the valuation of $x_i \in X$ in state s by $s(x_i)$. The central theme in our encoding of preference P as a Kripke structure $K(P)$ relies on the following two mappings between $\delta(P)$ to $K(P)$. *First*, for each node $\gamma \in A$ in $\delta(P) = G(A, E)$, there is a corresponding state $s^\gamma \in S$ in $K(P)$ such that for all preference variables $x_i \in X$, $\gamma(x_i) = s^\gamma(x_i)$. Thus there are a total $|S| = |A| = 2^{|X|}$ number of states in $\delta(P)$ and $K(P)$. *Second*, the transition relation T in $K(P)$ is such that paths in $K(P)$ correspond to paths in $\delta(P)$.

Recall from the syntax and semantics of various preference languages (Chapter 2) that for each preference statement p in P where the preference over a variable x_i is specified, whether or not p induces a flip from one outcome (node in $\delta(P)$) to another in $\delta(P)$ is constrained by sets of conditions that ensure:

(a) valuations of x_i in the source node is worse than that in the destination node (improving flip);

(b) valuations of parent variables of x_i (variables on whose values the improvement of x_i as per p depends) in the source and destination nodes equal their valuations in p; and

(c) all variables other than x_i and those that are less important than x_i remain unchanged in the destination node, according to the *ceteris paribus* principle.

Analogous to the definition of flips in $\delta(P)$, we define transitions between two states in $K(P)$ in terms of enabling conditions on the state variables x_i in source and destination states. According to the semantics of the input language of the model checker NuSMV, by default all the state variables in the model are allowed to change non-deterministically, unless explicitly constrained by a guard condition. Hence, to be able to restrict transitions other than those satisfying the above conditions, we use another set H of auxiliary variables h_is (corresponding to the x_is) to label the transitions between states:

$$h_i = \begin{cases} 0 & \Rightarrow & \text{value of } x_i \text{ must not change in a} \\ & & \text{transition in the Kripke structure } K(P) \\ 1 & \Rightarrow & \text{otherwise} \end{cases} \quad (1)$$

In the above definition, the valuation $h_i = 0$ on a transition restricts x_i from changing in the transition; and the valuation $h_i = 1$ allows x_i to (possibly) change in the transition. Hence, by setting auxiliary variables H to a combination of values, we can specify *guard conditions* for x_i such that the transitions allowed in $K(P)$ between states s and s' (that correspond to nodes γ and γ' in $\delta(P)$) are those that precisely correspond to the edges in $\delta(P)$ between γ and γ'.

Translating Preference Statements to Guard Conditions. Given a preference statement p in a specification P, we show how to generate the set of guards that define transition relations in $K(P)$ corresponding to the (improving) flips in $\delta(P)$. The guard condition $\mathcal{G}(p)$ will consist of two parts: (a) a constraint $Allow(p)$ on the assignment of X (including any conditional dependency constraint ϱ mentioned in p) that must be satisfied in the current state for any outgoing transition induced by p;[1] and (b) a constraint $Restrict(p)$ on H (in the current state) which specifies (by the

---

[1]We say "transition induced by p" to refer a transition that corresponds to an improving flip induced by p.

semantics above) which of the x_js must not change in a transition induced by p. In the following, we define $Allow(p)$ and $Restrict(p)$ for the various types of preference statements.

i. *Guards for Statements of Intra-attribute Preference.*   For each preference statement p of the form $\varrho : x_i = v_i \succ_{x_i} x_i = v_i'$ in a CP-net (see syntax of CP-nets in Section 2.3 of Chapter 2), the guard condition with respect to p is as follows.

$$Allow(p) \quad := \quad \varrho \wedge x_i = v_i' \wedge h_i = 1$$
$$Restrict(p) \quad := \quad \bigwedge_{x_j \in X \setminus \{x_i\}} h_j = 0$$

In the above, as per $Allow(p)$, x_i can change in any outgoing transition from the current state in $K(P)$ ($h_i = 1$ in the current state) whenever the condition ϱ in p is satisfied and the valuation of x_i has a less preferred valuation according to p in the current state ($x_i = v_i'$). All other variables x_j remain unchanged in any such transition from the current state as per $Restrict(p)$.

Example 4.12   Consider the cyberdefense CP-net example in Section 2.3, the preference statement p_2, i.e.,

$$\text{true} : A = \text{Low} \succ_A A = \text{High}$$

induces four improving flips from α to β, i.e., whenever $\alpha(A) = \text{High}$, $\beta(A) = \text{Low}$, $\alpha(E) = \beta(E)$, and $\alpha(F) = \beta(F)$. Note that here $\varrho = \text{true}$ because the preference is unconditional, i.e., it holds always, as long as the *ceteris paribus* condition on other variables holds.

This is succinctly encoded as a guard within NuSMV's input modeling language; Figure 4.2 shows part of a NuSMV model that encodes this preference statement. Note that the preference variables lower cased and auxiliary variables are named ha instead of h_a etc. so as to abide by NuSMV variable naming conventions. More information on the NuSMV modeling language constructs used is detailed later in the chapter.

```
next(a) :=
  case
    a=High & ha=1 & he=0 & hf=0 : Low;
    TRUE : a;
  esac;
```

Figure 4.2: Encoding of guard for intravariable preference p_2 in CP-net P^{CP} in the input language of NuSMV model checker.

Similarly, Figure 4.3 shows the encoding of guard in NuSMV for p_3 in P^{CP}

$$E = \text{Code-Available} : F = \text{Fix-Unavailable} \succ_F F = \text{Fix-Available}$$

which is a preference on F conditioned on the value of E.

```
next(f) :=
  case
    e=Code-Available & f=Fix-Available & ha=0 & he=0 & hf=1 : Fix-Unavailable;
    TRUE : f;
  esac;
```

Figure 4.3: Encoding of guard for intravariable preference p_3 in CP-net P^{CP} in the input language of NuSMV model checker.

ii. *Guards for Statements of Relative Importance of an Attribute over a Set of Attributes.*

Let p be a preference statement in a CP-theory of the form $\varrho : x_i = v_i \succ_{\{x_i\}} x_i = v_i'$ $[\Omega]$, where $\varrho = \rho(x_i)$. The set Ω of variables that are relatively less important than x_i (as per p) must be allowed to take any value or change unrestrictedly simultaneously in any transition where x_i flips to an improving valuation. In order to enable the simultaneous change in the values of x_i and each $x_j \in \Omega$ in the same transition, we specify appropriate guard conditions based on the valuations of the variable x_i and that of each of the variables in the set Ω as follows.

$$Allow(p) \quad := \quad \varrho \wedge x_i = v_i' \wedge h_i = 1 \wedge \left(\bigwedge_{x_j \in \Omega} h_j = 1 \right)$$
$$Restrict(p) \quad := \quad \bigwedge_{x_k \in X \setminus \{x_i\} \setminus \Omega} h_k = 0$$

In the above, $\left(\bigwedge_{x_j \in \Omega} h_j = 1 \right)$ is the additional term from the previous case (for intra-variable preference), which allows all variables less important than x_i to be changed to any value as long as the value of variable x_i can be improved. This models relative importance of x_i over x_j for all $x_j \in \Omega$. The above guard conditions also give the guard condition for a TCP-net preference statement, as TCP-nets are a special case of CP-theories with $|\Omega| \leq 1$.

Example 4.13 Consider in the cyberdefense TCP-net P^{TCP} (see Section 2.4) the preference statement p_2', that expressed the relative importance of A over E in addition to the intra-variable preference over A as in P^{CP}.

$$\text{true} : A = \text{Low} \succ_A A = \text{High} \ [\{E\}]$$

As p_2' specifies relative importance of A over E, it induces additional improving flips from α to β whenever $\alpha(A) = \text{High}$, $\beta(A) = Low$, and $\alpha(F) = \beta(F)$, without requiring $\alpha(E) =$

$\beta(E)$. Compare this with the preference statement p_2 (see Section 2.3), which does not have any relative importance constraint. Both p_2 and p_2' are unconditional preference; as a result $\varrho = \text{true}$. Figure 4.4 shows part of a NuSMV model that encodes p_2' (compare this with the encoding of p_2 in Figure 4.2). The guards for the change in the values of e involving ha=1 & hb=1 enable the flips to be induced in $\delta(P^{TCP})$ due to p_2'. Note that, in addition to the guarded transitions for e that correspond to modeling p_2', there can be other guarded transitions for e. For instance, the preference statement p_1 (see Section 2.3) describes intra-variable preference over the preference variable E, which will result in a guarded transition in NuSMV:

```
next(e) :=
   case
      . . .
      e=Code-Unavailable & ha=0 & he=1 & hf=0 : Code-Available;
                                   -- intravariable preference.
      . . .
   esac;
```

```
next(a) :=
   case
      a=High & ha=1 & he=1 & hf=0 : Low; -- intravariable preference.
      TRUE : a;
   esac;
next(e) :=
   case
      a=High & ha=1 & he=1 & hf=0 : {Code-Available,Code-Unavailable};  -- relative
   importance.
      . . .
      TRUE : e;
   esac;
```

Figure 4.4: Encoding of guard for relative importance preference p_2' in TCP-net P^{TCP} in the input language of NuSMV model checker.

iii. *Guards for Statements of Relative Importance over Arbitrary Sets of Attributes.* The improving flip induced by a preference statement p of the form $\Phi^+, \Phi^- : \Psi_1 \succ \Psi_2$ will ensure all variables in the set Ψ_1 are improved and those in Ψ_2 are worsened (according to the CI-net semantics, this amounts to trading off elements in Ψ_2 for those in Ψ_1) in a transition, while all other variables remain unchanged.

Let $\varrho = (\bigwedge_{x_i \in \Phi^+} x_i = 1) \wedge (\bigwedge_{x_j \in \Phi^-} x_j = 0)$.

$$Allow(p) \ := \ \varrho \wedge (\textstyle\bigwedge_{x_i \in \Psi_1} x_i = 0) \wedge (\textstyle\bigwedge_{x_j \in \Psi_2} x_j = 1)$$
$$\wedge \ (\textstyle\bigwedge_{x_k \in \Psi_i \cup \Psi_2} h_k = 1)$$
$$Restrict(p) \ := \ \textstyle\bigwedge_{x_j \in X \setminus (\Psi_1 \cup \Psi_2)} h_j = 0$$

The SMV model encoding the above guard conditions will follow a similar style to the CP-net and TCP-net example given previously. We leave it as an exercise to the reader to apply the above rules for the CI-net example in Example 2.5.

The Transition Relation of $K(P)$. Given a preference statement p in any of the *ceteris paribus* languages, the guard condition that constrains changes to the valuation of preference variables in any state in $K(P)$ is given by $\mathcal{G}(p) = Allow(p) \wedge Restrict(p)$. A state $s \in S$ in $K(P)$ is said to satisfy the guard condition $\mathcal{G}(p)$, denoted $s \models \mathcal{G}(p)$, if and only if it respects the constraints on all the x_i's and h_i's imposed by $\mathcal{G}(p)$; and this implies the possibility of existence of outgoing transitions from s corresponding to improving flips in $\delta(P)$ induced by p. $K(P)$'s transitions are modeled as a composition of all the guard conditions induced by the corresponding preference statements of P.

The overall transition relation T of $K(P)$ is specified as the union of the transitions[2] induced by all the preference statements in P.

Figures 4.5, 4.6, and 4.7 show the Kripke structures $K(P^{CP})$, $K(P^{TCP})$, and $K(P^{CPTh})$ constructed from the CP-net P^{CP} (see Section 2.3 and Figure 2.2) and TCP-net P^{TCP} (see Section 2.4, Figure 2.3) and CP-theory P^{CPTh} (see Section 2.5 and Figure 2.4), respectively, according to the above encoding.

The states are annotated with the values of the preference (or state) variables in the Kripke structure. The transitions are annotated with the values of the boolean variables in H, denoting which state variables must remain unchanged and which ones may change due to the transition. The model checker non-deterministically sets the value of variables in H; as a result each state has $2^{|H|}$ (corresponding to all possible combinations of values for variables in H) outgoing transitions (in our example, $|H| = 3$). Note that some combinations of these values do not satisfy the guard conditions corresponding to improving flip; for those combinations of values of variables in H, the transitions result in self-loops. For clarification of the diagrams, we have omitted the self-loops.

2.2   CORRECTNESS OF THE CONSTRUCTION OF $K(P)$

Note that the semantics of each preference statement in P is directly encoded as a set of guarded transitions in $K(P)$. If there is a valid flip from γ to γ' in $\delta(P)$, then there is a transition from a corresponding state s to another state s' in $K(P)$. In other words, the guard conditions enable only the transitions in $K(P)$ that correspond to valid flips in $\delta(P)$. Hence, the correctness of the construction of $K(P)$ with respect to modeling the induced preference graph $\delta(P)$ can be proved

---

[2]Note that a state s in $K(P)$ may also contain transitions to itself (self-transitions) because $h_i = 1$ does not necessarily imply that x_i will change; rather, it allows non-deterministic choice for the valuation of x_i.

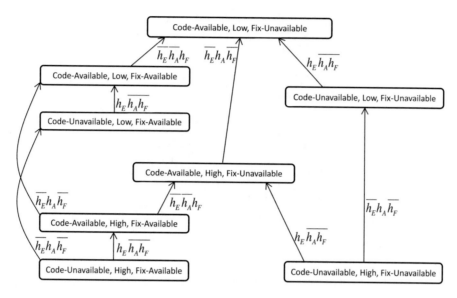

Figure 4.5: $K(P^{CP})$: Kripke structure encoding the semantics of P^{CP}.

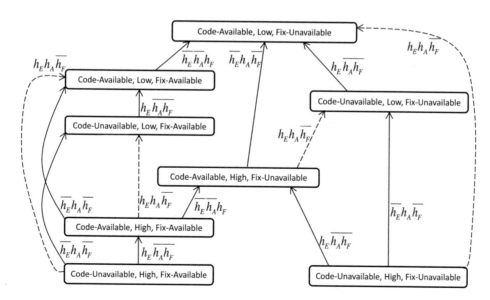

Figure 4.6: $K(P^{TCP})$: Kripke structure encoding the semantics of P^{TCP}.

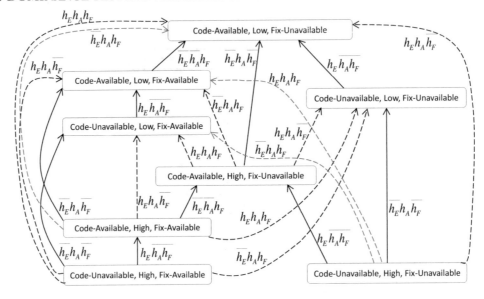

Figure 4.7: $K(P^{CPTh})$: Kripke structure encoding the semantics of P^{CPTh}.

by the construction outlined so far or by contradiction. We simply formally state the main result and leave the proof to be verified by the reader.

Dominance testing with respect to P is equivalent to reachability over $\delta(P)$, i.e., $\gamma' \succ \gamma$ with respect to P if and only if there is a path from γ to γ' in $\delta(P)$ (see Definition 2.16). Therefore, to prove the correctness of the construction of $K(P)$, we show that there is a path (flipping sequence) from γ to γ' in $\delta(P)$ if and only if there is a corresponding sequence of transitions from a state $s \in S^{\gamma}$ to a state $s' \in S^{\gamma'}$ in $K(P)$. This will ensure that the state space explored by the model checker when given $K(P)$ as input corresponds to $\delta(P)$.

Let $\gamma \xrightarrow{p} \gamma'$ denote that p induces a flip from γ to γ', $s \to s'$ denote that a transition exists from state s to a state s' in $K(P)$, and $s \xrightarrow{\star} s'$ denote that a sequence of transitions exists in $K(P)$ from s to s'.

Theorem 4.14 (Correctness of Kripke Encoding) *Let $P = \{p_1, p_2 \ldots\}$ be a preference specification over an outcome space $\mathcal{O}$, and the corresponding Kripke structure $K(P) = \langle S, S_0, T, L \rangle$ resulting from the above encoding. Then*

$$\forall \gamma_1, \gamma_2 \ldots \gamma_n \in \mathcal{O} : \left[\left(\forall i \in [1, n-1] : \gamma_i \neq \gamma_{i+1} \right) \wedge \left(\gamma_1 \xrightarrow{p_1} \gamma_2 \xrightarrow{p_2} \ldots \xrightarrow{p_{n-1}} \gamma_n \right) \right]$$
$$\Leftrightarrow \left[\exists s_1, s_2, \ldots s_n \in S : \left(\forall j : s_j \in S^j \right) \wedge \left(s_1 \xrightarrow{\star} s_2 \xrightarrow{\star} \ldots \xrightarrow{\star} s_n \right) \right]$$

The above theorem implies that the induced preference graph $\delta(P)$ and $K(P)$ are equivalent with respect to reachability between any pair of distinct outcomes, i.e., they model the same set of paths between all pairs of outcomes. The next question is then how does one use $K(P)$ to answer dominance (reachability) with respect to P without constructing $\delta(P)$.

3   ANSWERING DOMINANCE QUERIES VIA MODEL CHECKING

We have presented the encoding of preference statements as guarded transition relations in NuSMV model checker. The complete listings of the Kripke structure encodings for some of the example CP-nets, TCP-nets, CP-theories, and CI-nets are available in the Appendix A. We now proceed to show how dominance queries with respect to a given preference specification can be described as temporal queries in the CTL Logic (see Chapter 3).

3.1   VERIFYING DOMINANCE

Given a Kripke structure $K(P)$ that encodes the induced preference graph of a preference specification P, determining whether α dominates β with respect to P amounts to verifying whether the states in $K(P)$ corresponding β can reach the states corresponding to α. The CTL property that describes this reachability pattern is

$$\varphi_\beta \rightarrow \mathbf{EF}\varphi_\alpha$$

The formula φ_α and φ_β denote the conjunction of valuation of (preference) variables that are true in the outcome α and β, respectively. Therefore, the Kripke structure states, corresponding to α and β, satisfy φ_α and φ_β, respectively.

A state in the Kripke structure $K(P)$ is said to satisfy the above formula if and only if when the state satisfies φ_β (i.e., valuations of variables of X in that state correspond to those in β), there exists a path or a sequence of transitions $s^\beta = s_1 \rightarrow s_2 \rightarrow \cdots \rightarrow s_n = s^\alpha$ (where $\forall x_i \in X : s^\alpha(x_i) = \alpha(x_i)$ and $s^\beta(x_i) = \beta(x_i)$) such that $n > 1$. In short, a state in the Kripke structure $K(P)$ corresponding to $\delta(P)$ satisfies the above CTL formula if and only if α dominates β with respect to P (Theorem 4.14). We use the model checker NuSMV to verify the satisfiability of a CTL formula $\varphi_\beta \rightarrow \mathbf{EF}\varphi_\alpha$.

Example 4.15   For the TCP-net P^{TCP} in Figure 2.1(b), the dominance of $\alpha = \langle$Code-Unavailable, Low, Fix-Unavailable$\rangle$ over $\beta = \langle$Code-Available, High, Fix-Available$\rangle$ corresponds to the satisfiability of the CTL formula

$$\varphi : (\mathtt{A} = \text{High} \wedge \mathtt{E} = \text{Code-Available} \wedge \mathtt{F} = \text{Fix-Available}) \rightarrow$$
$$\mathbf{EF}(\mathtt{A} = \text{Low} \wedge \mathtt{E} = \text{Code-Unavailable} \wedge \mathtt{F} = \text{Fix-Unavailable})).$$

Note that NuSMV asserts that a formula φ is verified only if *every* initial state satisfies φ. Therefore, we initialize x_i to $\alpha(x_i)$ for $x_i \in \{A, E, F\}$ to restrict the start states to those corresponding to the outcome α in the encoded Kripke structure. We also initialize all the change variables h_i to 0, so that transitions corresponding to all possible improving flips from α in the Kripke model are explored by the model checker non-deterministically.

In NuSMV, the satisfiability of φ can be verified by the specification SPEC φ, and the verification returns 'true' in our example, thereby establishing that the dominance holds.

3.2   EXTRACTING A PROOF OF DOMINANCE

We can use the NuSMV model checker to obtain a proof that an alternative α dominates another alternative β (i.e., an improving flipping sequence from β to α exists) as follows. Suppose α dominates β. This implies that the CTL formula $\varrho : \varphi_\beta \rightarrow \mathbf{EF}\varphi_\alpha$ holds, where φ_β and φ_α correspond to the boolean formulas encoding the respective outcomes β and α. Hence, in this case if we provide the formula $\neg\varrho$ (i.e., $\neg(\varphi_\beta \rightarrow \mathbf{EF}\varphi_\alpha)$) as input to the model checker, the model checker will return 'false', and provide us with the sequence of states (as below) corresponding to the improving flipping sequence from β to α. In our example, $\alpha = \langle$Code-Unavailable, Low, Fix-Unavailable$\rangle$ dominates $\beta = \langle$Code-Available, High, Fix-Available$\rangle$.

As a result, for the property expressed using the following

SPEC $\neg$(A = High $\wedge$ E = Code-Available $\wedge$ F = Fix-Available) $\rightarrow$
$$\mathbf{EF}(A = \text{Low} \wedge E = \text{Code-Unavailable} \wedge F = \text{Fix-Unavailable})$$

NuSMV returns the sequence of outcomes

$\langle$Code-Available, High, Fix-Available$\rangle$ $\rightarrow$
$\langle$Code-Available, High, Fix-Unavailable$\rangle$ $\rightarrow$
$\langle$Code-Unavailable, Low, Fix-Unavailable$\rangle$

as shown in Figure 4.8, which is a proof, i.e., improving flipping sequence verifying the dominance.

Figure 4.8 shows the output from the NuSMV model checker for the above CTL query. Observe that the transition from state 1.1 to 1.2 is effected by the semantics of the preference statement p_3; and that from state 1.2 to 1.3 is effected by the semantics of the preference statement p_2'.

The same approach as described above can be used to answer dominance queries for any of the *ceteris paribus* preference languages such as CP-nets, CI-nets, and CP-theories. The sample SMV models, CTL formulas for dominance queries, and the corresponding NuSMV output traces are provided in Appendix A. An interested reader will see all the techniques discussed in this chapter applied in those listings.

```
-- specification
(
  (
    (a = High & e = Code-Available) & f = Fix-Available)
         -> ! (EX (EF ((a = Low & e = Code-Unavailable) & f = Fix-Unavailable))
  )
) is false
-- as demonstrated by the following execution sequence
Trace Description: CTL Counterexample
Trace Type: Counterexample.
-> State: 1.1 <-
  f = Fix-Available
  a = High
  e = Code-Available
  start = TRUE
-> Input: 1.2 <-
  ha = 0
  he = 0
  hf = 1
-> State: 1.2 <-
  f = Fix-Unavailable
-> Input: 1.3 <-
  ha = 1
  he = 1
  hf = 0
-> State: 1.3 <-
  a = Low
  e = Code-Unavailable
```

Figure 4.8: Output from NuSMV model checker on querying the Kripke model with the CTL formula for the negation of dominance.

3.3   SUMMARY AND DISCUSSION

We have described the first practical solution to the problem of determining whether an alternative dominates another with respect to a set of qualitative preferences. Our approach relies on a reduction of the dominance testing problem to reachability analysis in a graph of alternatives. We have provided an encoding of CP-nets, TCP-nets, and other languages in the form of a Kripke structure for CTL.

We have shown how to: (a) directly and succinctly encode preference semantics as a Kripke structure; (b) compute dominance by verifying CTL temporal properties against this Kripke structure; and (c) generate a proof of dominance. We have shown how to compute dominance using

NuSMV, a model checker for CTL. This approach to dominance testing via model checking allows us to take advantage of continuing advances in model-checking.

Although our examples focused on TCP-nets, our approach can be applied to any preference language for which the semantics is given in terms of the satisfiability of graph properties (including preference specifications that induce cyclic preferences). The presented approach can be extended for other reasoning tasks such as finding whether a given alternative is the *least* (*or most*) *preferred* among all the alternatives.

Although we have used the NuSMV model checker in this chapter, any model checker that accepts a Kripke structure as input can be used to realize our approach to dominance testing. Hence, it should be possible to take advantage of specialized techniques that have recently been developed to improve the performance of model checkers [9, 25, 33, 53, 92].

CHAPTER 5

Verifying Preference Equivalence and Subsumption

In the previous chapter, we presented an encoding of preference semantics (induced preference graph) as a system specification and have discussed the use of model checking to test whether one alternative dominates another with respect to a given set of preferences. In this chapter we consider the problem of determining whether one set of preference statements is *equivalent* to another or whether one set of preferences *subsumes* another. Preference equivalence and preference subsumption testing find application in determining the substitutability of preference profiles [46] and in stable matching problems [43]. Other potential applications include matching or finding users with similar preferences in recommender systems [88] and checking (dis)agreement of preferences of a set of agents in group decision making [77].

We will extend the encoding strategy presented in the previous chapter and show how model checking can be used to determine preference equivalence and subsumption in this chapter. Our approach can be summarized as follows. Given the objective of verifying the preference equivalence between the preference specifications P and P', in order to ensure that a dominance is satisfied in P if and only if it is satisfied in P', we encode P and the inverse[1] of P' as one Kripke structure. Then we model check this Kripke structure against a CTL property which essentially states that if a state s (i.e., an outcome) can reach s' using the transitions resulting from P then s' can reach s as per P', and vice versa. Being based on model checking, the technique has two primary advantages. First, for the preferences that are not equivalent, the counterexamples generated by model checker as a result of violation of CTL property can directly present the proof of non-equivalence; the proof includes some dominance relation satisfied in one preference specification and not in the other. Second, the encoding does not require that the preferences, whose equivalence is to be verified, are expressed in the same language.

Because things get too complicated with modeling two sets of preferences in one data structure, we pick two simple CP-nets as running examples to discuss our approach to preference equivalence testing and preference subsumption testing, instead of the examples introduced in the first chapter. We provide worked out examples of this approach for the cyberdefense example introduced in Chapter 1 in an appendix. We note that while the running examples in this chapter are simple CP-nets, the approach presented here is applicable TCP-nets, CI-nets, and CP-theories as well.

---

[1]α dominates β in P' if and only if β dominates α in the inverse of P'.

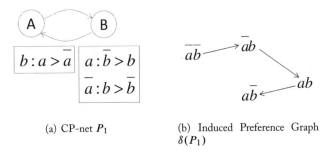

(a) CP-net P_1

(b) Induced Preference Graph $\delta(P_1)$

Figure 5.1: Example preference specification: CP-net P_1.

1   PREFERENCE EQUIVALENCE AND PREFERENCE SUBSUMPTION

We now turn to the problem of verifying the preference equivalence and preference subsumption relationships between two sets of preferences. We first formally define preference equivalence and preference subsumption.

Definition 5.1   Preference Equivalence and Preference Subsumption.   Let P_1 and P_2 be two CP-nets over a set of variables V. Let $\succ^{1\star}$ and $\succ^{2\star}$ represent the transitive closures of the preference relations $\succ^1$ and $\succ^2$ induced by P_1 and P_2 respectively over the set of alternatives.

  i. P_1 is said to **preference subsume** P_2, denoted $P_1 \sqsupseteq P_2$ or $P_2 \sqsubseteq P_1$, iff $\forall \gamma, \gamma' : \gamma \succ^2 \gamma' \Rightarrow \gamma \succ^1 \gamma'$, or equivalently $\succ^{1\star} \sqsupseteq \succ^{2\star}$.

  ii. P_1 is said to be **preference equivalent** to P_2, denoted $P_1 \equiv P_2$ iff $P_1 \sqsupseteq P_2 \wedge P_2 \sqsupseteq P_1$.

In the above, equivalence and subsumption of P_1 and P_2 are defined in terms of the transitive closures of the respective induced preferences, namely $\succ^{1\star}$ and $\succ^{2\star}$ that represent the set of all improving flipping sequences for P_1 and P_2, and not simply in terms of $\succ^1$ and $\succ^2$ that represent only the set of improving flips induced by the respective preference statements in P_1 and P_2. This is necessary because the dominance relation is transitive.

In other words, verifying the semantic equivalence of two sets P_1 and P_2 of preference statements amounts to checking that for each *edge* from γ to γ' in $\delta(P_1)$, there exists a corresponding *path* from γ to γ' in $\delta(P_2)$ and vice-versa. It is worth noting here that the above holds for any preference language that has a flipping-sequence based (*ceteris paribus*) semantics. Because dominance testing is PSPACE-complete for CP-nets [16], TCP-nets [18], CP-theories [93], and CI-nets [17], preference equivalence testing (and preference subsumption testing) are arguably PSPACE-complete for these preference languages.

Example 5.2   Figure 5.1 shows a CP-net P_1 and its induced preference graph $\delta(P_1)$. Figure 5.2 shows a CP-net P_2 with its induced preference graph $\delta(P_2)$. From the preference statements of P_1

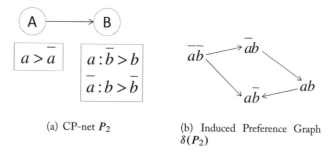

(a) CP-net P_2 (b) Induced Preference Graph $\delta(P_2)$

Figure 5.2: Example preference specification: CP-net P_2.

and P_2, it may appear that P_2 subsumes P_1, i.e., $P_1 \sqsubseteq P_2$, because the only difference between them is the preference over variable A: in P_1 it is conditioned on $b = 1$, whereas in P_2 it is unconditional. On the other hand, it may not be as intuitive to conclude that $P_2 \sqsubseteq P_1$. However, this is indeed the case: $P_1 \sqsubseteq P_2$ and $P_2 \sqsubseteq P_1$, i.e., $P_1 \equiv P_2$, because the induced preference graphs $\delta(P_1)$ and $\delta(P_2)$ are equivalent in terms of the reachability between any pair of alternatives. The unconditional preference over A in P_2 gives rise to an additional edge in $\delta(P_2)$ from $\bar{a}\bar{b}$ to $a\bar{b}$ that has no corresponding edge in $\delta(P_1)$, but the same has a corresponding path in $\delta(P_1)$: $\bar{a}\bar{b} \rightarrow \bar{a}b \rightarrow ab \rightarrow a\bar{b}$, which makes $\delta(P_1)$ and $\delta(P_2)$ (and hence P_1 and P_2) equivalent.

2   DATA STRUCTURES TO REPRESENT SEMANTICS OF TWO SETS OF PREFERENCE

As we have outlined before, the central theme of our technique for equivalence and subsumption verification is to combine the preferences induced by one preference specification with the inverse of the other. The following sections detail the steps.

2.1   INVERSE INDUCED PREFERENCE GRAPH

We begin with defining the inverse of an induced preference graph with respect to a preference specification P, which represents the inverse of the preference relation induced by P on the set of outcomes.

Definition 5.3   Inverse Induced Preference Graph.   Given a preference specification P, the *inverse induced preference graph* $\delta^-(P)$ is constructed by generating $\delta(P)$ and reversing the direction of all the edges.

Example 5.4   Figure 5.3(a) and (b) show the inverse induced preference graphs $\delta^-(P_1)$ and $\delta^-(P_2)$ for the CP-nets P_1 and P_2 shown in Figures 5.1 and 5.2 respectively.

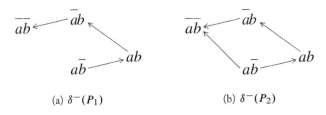

(a) $\delta^-(P_1)$ (b) $\delta^-(P_2)$

Figure 5.3: Inverse induced preference graphs for CP-nets P_1 and P_2.

2.2   COMBINED INDUCED PREFERENCE GRAPH

Given two sets of preference statements P_1 and P_2, we first consider preference subsumption test-ing, i.e., verifying whether $P_1 \sqsubseteq P_2$, since preference equivalence testing, i.e., checking whether $P_1 \equiv P_2$ amounts to verifying whether $P_1 \sqsubseteq P_2 \wedge P_2 \sqsubseteq P_1$. Verifying $P_1 \sqsubseteq P_2$ amounts to ver-ifying that for each edge (γ, γ') in $\delta(P_1)$ there is a corresponding path from (γ, γ') in $\delta(P_2)$. This can be reduced to verifying a reachability property in a graph, namely the *combined induced pref-erence graph* of P_1 and P_2, denoted $\delta(P_1, P_2)$, that embeds the semantics of P_1 and the inverse of the semantics of P_2.

Definition 5.5   Combined Induced Preference Graph.   Given two sets P_1 and P_2 of preference statements over a set of variables V, the *combined induced preference graph* $\delta(P_1, P_2)$ is a directed graph $G(N, E)$ with a labeling function L that is constructed as follows. The nodes N correspond to the set of alternatives generated by V. There is an edge $e_{\gamma,\gamma'} = (\gamma, \gamma') \in E$ if and only if there is an edge from γ to γ' in $\delta(P_1)$ or $\delta^-(P_2)$, and it is associated with a label

$$\mathcal{L}(e_{\gamma,\gamma'}) = \begin{cases} \{1\} & \text{if } \gamma' \succ^1 \gamma \text{ and } \gamma \not\succ^2 \gamma' \\ \{2\} & \text{if } \gamma \succ^2 \gamma' \text{ and } \gamma' \not\succ^1 \gamma \quad (1) \\ \{1,2\} & \text{if } \gamma' \succ^1 \gamma \text{ and } \gamma \succ^2 \gamma' \end{cases}$$

Note that in the graph $\delta(P_1, P_2)$, each edge $(\gamma, \gamma') \in E$ corresponds to an improving flip from γ to γ' induced by P_1, a worsening flip from γ to γ' induced by P_2, or both. Figure 5.4 shows the combined induced preference graph $\delta(P_1, P_2)$ with respect to the CP-nets P_1 and P_2 shown in Figures 5.1 and 5.2 respectively, consisting of the edges in $\delta(P_1)$ (solid arrows, labeled 1) and those in $\delta^-(P_2)$ (dotted arrows, labeled 2).

Recall that verifying $P_1 \sqsubseteq P_2$ is equivalent to verifying that for each edge (γ, γ') in $\delta(P_1)$ there exists a corresponding path from (γ, γ') in $\delta(P_2)$, or in other words, there exists a corre-

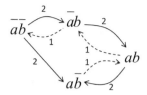

Figure 5.4: Combined induced preference graph $\delta(P_1, P_2)$.

sponding path from (γ', γ) in $\delta^-(P_2)$. Therefore, in terms of the combined induced preference graph $\delta(P_1, P_2)$, the following proposition can be stated.

Proposition 5.6   Necessary and Sufficient Condition for Subsumption $P_1 \sqsubseteq P_2$ *holds if and only if: for each edge from γ to γ' in $\delta(P_1, P_2)$ that includes label $\{1\}$, there exists a path from γ' to γ such that each edge in the path includes the label $\{2\}$.*

The above forms the basis of our model checking based approach to preference subsumption testing.

3   KRIPKE STRUCTURE ENCODING FOR PREFERENCE EQUIVALENCE AND SUBSUMPTION

In order to verify $P_1 \sqsubseteq P_2$, we first construct a Kripke structure $K(P_1, P_2)$ that encodes the combined induced preference graph $\delta(P_1, P_2)$. To achieve this, in the next subsection we will recall modeling the induced preference graph of a single preference specification as a Kripke structure, and then extend it in the following subsection to be able to represent two preference specifications in the same model.

3.1   MODELING OF PREFERENCE SEMANTICS: EXTENSION FOR PREFERENCE EQUIVALENCE AND PREFERENCE SUBSUMPTION REASONING

Recall from Chapter 4, the improving flips as per preference statements are encoded as guarded transition relations in a Kripke structure. Each state in the Kripke structure is identified by the valuations of preference variables X, i.e., each state corresponds to an outcome. The guards on the transitions correspond to the valuations of the variables in the source states and a set H of boolean variables. The variables in H decided whether or not a preference variable value can change as a result of the transition, and the variables in H are non-deterministically set by the model checker. The encoding described in Chapter 4 ensured that for every sequence of improving flips in the induced preference graph $\delta(P)$ there is a sequence of transitions in the corresponding Kripke structure $K(P)$, which allowed us to verify dominance by in turn verifying reachability in the $K(P)$. Note that there are transitions in $K(P)$ that result in self-loops because the value of

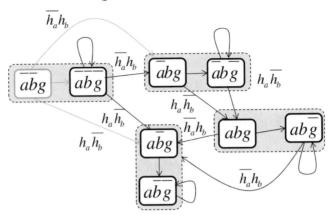

Figure 5.5: $K(P_1)$: Kripke encoding for the semantics of P_1.

variables in H (non-deterministically set by the model checker) does not satisfy the guards for improving flips; i.e., these self-loop transitions do not correspond to any improving flips.

For the purpose of verifying equivalence and subsumption checking, we will need a new boolean variable g (*global change variable*) to annotate the states in the Kripke structure. For every transition between states s and t for which there exists some preference variable x such that $s(x) \neq t(x)$, the value of g in t is set to 1. In other words, states s and t correspond to different outcomes in the induced preference graph, and these transitions correspond to improving flips between them. For all other transitions, the value of g is set to 0—these transitions do not correspond to improving flips. The introduction of the variable g changes the structure of $K(P)$, without violating the key correctness property that for every improving flipping sequence (of outcomes) in the $\delta(P)$ there is a corresponding transition sequence in $K(P)$.

Example 5.7   Figure 5.5 shows the Kripke structure $K(P_1)$ for the CP-net P_1 in Figure 5.1 with the newly introduced variable g. Note that due to the introduction of g, for each outcome, there exist two states in the Kripke—they only differ in the values of g; one with the g being true and other with g being false. These types of states are presented in the figure in grey boxes. For an outcome γ, let us denote s_g^γ as the state where g is true and $s_{\bar{g}}^\gamma$ as the state where g is false. Note that, for each set of values of variables in H that does not satisfy the conditions for improving flips, there is a transition from s_g^γ to $s_{\bar{g}}^\gamma$ and a self-loop on $s_{\bar{g}}^\gamma$. The transitions corresponding to improving flip connects the states corresponding to the worse outcome to the state corresponding to the better outcome where the value of g is set to true; this is presented as transitions with source grey box (representing that both states in the grey box are the source of the transition). For the sake of clarity, we only annotate the transitions with the values of variables in H whenever the transitions correspond to improving flip.

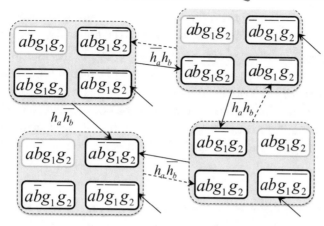

Figure 5.6: Kripke structure $K(P_1, P_2)$ encoding the combined induced preference graph $\delta(P_1, P_2)$.

3.2   ENCODING COMBINED INDUCED PREFERENCE GRAPH AS KRIPKE STRUCTURE

We verify preference subsumption by verifying the condition of Proposition 5.6 using a Kripke structure encoding $K(P_1, P_2)$ of $\delta(P_1, P_2)$. To achieve this, we must be able to distinguish the transitions in the model (corresponding to flips) induced by P_1 and P_2. We use two global change variables g_1 and g_2 to label the destination states of any transition as follows: $g_1 = 1$ if the state is reached via a transition induced by P_1 and $g_2 = 1$ if the state is reached via a transition induced by P_2 (more precisely the inverse of the given P_2, as described in the preceding subsection).

The variables g_1 and g_2 are modeled as *state* variables, and hence the states S of $K(P_1, P_2)$ are defined by the valuations of propositions $P_V = \{x_i | x_i \in V\} \cup \{g_1, g_2\}$. Hence, each outcome or alternative γ in $\delta(P_1, P_2)$ corresponds to a set $S^\gamma = \{s | s_{\downarrow V} = \gamma\}$ of states, where $s_{\downarrow V}$ denotes the *projection* of a state s described by P_V onto the set of variables $\{x_i | x_i \in V\} \subseteq P_V$. The variables g_1 and g_2 are initialized to 0 in $K(P_1, P_2)$, whereas x_i's are uninitialized, i.e., the model checker non-deterministically chooses and explores all possible combinations of assignments to x_i's. Hence the set of start states S_0 corresponds to the set of alternatives. The state space of $K(P_1, P_2)$ for the running example in this chapter is illustrated below.

Example 5.8   Figure 5.6 shows the Kripke structure $K(P_1, P_2)$ corresponding to $\delta(P_1, P_2)$ for our running example. The start states are marked with transitions without any source state. For clarity, the transitions between states corresponding to the same alternative are not marked, and the valuations of input variables a^0 and b^0 are not shown in the transitions. States corresponding to the same valuation of a and b are placed within dotted boxes. Further, transitions from the set of all states in a dotted box to the same destination state with a different valuation of a and b (in a different dotted box) are combinedly represented by a single arrow from dotted box to the destination state, e.g., the arrow from the dotted box containing the states corresponding to $\bar{a}b$ indicates the presence of transitions from all states in the box to $\bar{a}bg_1\bar{g}_2$.

Recall that the encoding of $\delta(P_1, P_2)$ as $K(P_1, P_2)$ is succinct in the sense that we do not explicitly specify each node and edge in $\delta(P_1, P_2)$ to construct the state space of $K(P_1, P_2)$; instead we simply specify the preference statements as guarded transition relations of Kripke structure and the model checker (using the power of non-determinism over the valuations of variables in H) automatically constructs and explores the Kripke structure state-space as needed (see Chapter 4 for details). The following holds by construction of $K(P_1, P_2)$ from $\delta(P_1, P_2)$ as just described.

Proposition 5.9   *Given CP-nets P_1 and P_2, and the Kripke structure $K(P_1, P_2) = \langle S, S_0, T, L \rangle$ (constructed from $\delta(P_1, P_2) = G(N, E)$ associated with labeling function $\mathcal{L}$),*

1. $\forall \gamma, \gamma', i \in \{1, 2\} : (\gamma, \gamma') \in E \wedge \mathcal{L}(e_{\gamma, \gamma'}) \supseteq \{i\}$ $\qquad \Rightarrow \exists s \to s' : s_{\downarrow V} = \gamma \wedge s'_{\downarrow V} = \gamma' \wedge s'(g_i) = 1.$

2. $\forall s, s' \in S : s \to s' \wedge s_{\downarrow V} \neq s'_{\downarrow V}$ $\qquad \Rightarrow \exists i \in \{1, 2\}, \gamma, \gamma' : s_{\downarrow V} = \gamma \wedge s'_{\downarrow V} = \gamma' \wedge (\gamma, \gamma') \in E \wedge \mathcal{L}(e_{\gamma, \gamma'}) \supseteq \{i\}.$

The full SMV source code listing for the Kripke structure $K(P_1, P_2)$ shown in Figure 5.6 is provided in Appendix C.

4   QUERYING $K(P_1, P_2)$ FOR SUBSUMPTION

We have already seen that verifying $P_1 \sqsubseteq P_2$ is equivalent to verifying that for each edge from γ to γ' in $\delta(P_1, P_2)$ that includes the label $\{1\}$, there exists a path from γ' to γ such that each edge in the path includes the label $\{2\}$ (Proposition 5.6). Because the set of start states S_0 in $K(P_1, P_2)$ corresponds to the set of alternatives in $\delta(P_1, P_2)$, the above reduces to verifying the following property in $K(P_1, P_2)$ by Proposition 5.9.

For each state $s \in S_0$ in $K(P_1, P_2)$, if there exists a transition to a state s' with $s'(g_1) = 1$, then there exists a path $s' = s_1 \to \ldots \to s_n \to s''$ in $K(P_1, P_2)$ such that $\forall 1 < i \leq n : s_i(g_2) = 1$ and $s_{\downarrow V} = s''_{\downarrow V}$.

Our objective is to express the above property in the language of Computation Tree Temporal Logic, CTL (Chapter 4), and automatically verify the temporal property with respect to $K(P_1, P_2)$ using a model checker such as NuSMV. One interesting and subtle challenge in realizing our objective stems from the fact that the condition requires checking the existence of paths starting from a state s and ending at a state s'' such that $s_{\downarrow V} = s''_{\downarrow V}$. However, CTL allows the specification and verification of temporal properties only with respect to states explored in the future, and therefore it is not possible to write a single temporal property in which the start state s corresponds to all possible outcomes in the corresponding induced preference graph.[2]

We address this challenge by introducing a set of *copy* variables, namely x_i^0's that are modeled in SMV model checker as *input* variables in $K(P_1, P_2)$ and hence not stored as part of the state. They are initialized with the valuations of the respective x_i's at the start of model exploration, and are constrained to remain invariant in the model; i.e., if the valuation of x_i at a start state $s \in S_0$ is v_i, then x_i^0 remains equal to v_i in all states along all paths starting from s. In other words, if the model checker begins exploration at state s, then the propositional formula $\psi = \bigwedge_i (x_i = x_i^0)$ can be used to refer to $s_{\downarrow V}$. Proceeding further, the following CTL formula encodes the condition for $P_1 \sqsubseteq P_2$.

$$\varphi : \mathbf{AX} \left(g_1 \Rightarrow \mathbf{EX} \ \mathbf{E} \left[g_2 \ \mathbf{U} \ (\psi \wedge g_2) \right] \right)$$

Recall from the CTL semantics described in Chapter 3 for $\mathbf{EX} \ \psi$, $\mathbf{AX} \ \psi$, and $\mathbf{E} \ [\psi_1 \mathbf{U} \ \psi_2]$. Therefore, φ holds in $K(P_1, P_2)$ whenever the following holds. For each transition $s \to s' \in T$ such that $s'(g_1) = 1$ (i.e., whenever $\mathbf{AX} \ g_1$ holds), there exists a transition $s' \to s''$ (i.e., $\mathbf{EX}$) such that s'' satisfies $\mathbf{E} \left[g_2 \ \mathbf{U} \ (\psi \wedge g_2) \right]$. That is, there is a path $s'' = s_1'' \to s_2'' \to s_k'' \ldots \to s_n''$ such that states till a state s_n'', where ψ also holds, is reached. Recall that propositional formula ψ is satisfied in states where the valuations of the preference variables are the same as those in the start state (denoted by s in this case; see above). Note that if there are no transitions $s \to s'$ such that $s'(g_1) = 1$, then φ trivially holds in s. While the formula φ seems to imply that only one particular improving flip in P_1 can be recovered using reverse flips in P_2, this is indeed without loss of generality. This is because if each improving flip in P_1 is matched by reverse paths from P_2, then the transitive closure of improving flips is also matched by reverse paths (using the transitive closure reverse paths corresponding to each improving flip). The following formalizes the fact that $K(P_1, P_2) \models \varphi$ actually corresponds to the subsumption $P_1 \sqsubseteq P_2$, and can be proved using Propositions 5.6 and 5.9.

Theorem 5.10 $K(P_1, P_2)$ *satisfies φ if and only if $P_1 \sqsubseteq P_2$.*

4.1   EXTRACTING A PROOF OF NON-SUBSUMPTION

The model checker returns `true` whenever φ is satisfied, i.e., $P_1 \sqsubseteq P_2$. Suppose that $P_1 \not\sqsubseteq P_2$. The model checker will then return `false`, and provide the justification/proof of unsatisfiability, essentially presenting a sequence that satisfies the negation of the φ (see above), which is:

$$\neg\varphi : \mathbf{EX} \left(g_1 \wedge \mathbf{AX} \ \neg\mathbf{E} \left[g_2 \ \mathbf{U} \ (\psi \wedge g_2) \right] \right)$$

---

[2]Note that it is necessary that s must be referenced without loss of generality, so that the single temporal property will force the model checker to consider all possible start states as precisely those corresponding to the set of outcomes. If s cannot be referenced without loss of generality, it will be necessary to pose an exponential number of queries to verify that the property holds for each pair of outcomes, which is practically infeasible.

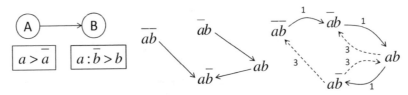

Figure 5.7: P_3 and the graphs $\delta(P_3)$ and $\delta(P_1, P_3)$.

The proof is presented in the form of a transition $s \to s'$ that corresponds to a flip from $\gamma = s_{\downarrow V}$ to $\gamma' = s'_{\downarrow V}$ such that (a) s' satisfies g_1 (implying that there is a path from s to s' as per P_1) and (b) s' satisfies $\mathtt{AX} \ \neg\mathtt{E}[\ g_2 \ \mathtt{U} \ (\psi \ \wedge \ g_2)\ ]$ (implying that there is no path from s' back to any state s'' with $s'' \downarrow_V = s \downarrow_V$ as per P_2). In other words, the transition $s \to s'$ corresponds to a flip from γ to γ' induced by P_1 but not by P_2.

4.2   VERIFYING PREFERENCE EQUIVALENCE

$P_1 \equiv P_2$ can be computed by verifying both φ and the following formula in $K(P_1, P_2)$.

$$\varphi' : \mathtt{AX} \left(g_2 \Rightarrow \mathtt{EX} \ \mathtt{E}\big[g_1 \ \mathtt{U} \ (\psi \ \wedge \ g_1)\big] \right)$$

Note that φ' verifies $P_2 \sqsubseteq P_1$. Hence by Definition 5.1, $P_1 \equiv P_2$ iff $\varphi \wedge \varphi'$ is verified in $K(P_1, P_2)$.

In our running example (Figure 5.4), the formula $\varphi \wedge \varphi'$ is verified in $K(P_1, P_2)$, proving that $P_1 \equiv P_2$. Now consider another CP-net P_3 and its relationship with P_1 shown in Figure 5.7. Note that φ' is verified in $K(P_1, P_3)$, i.e., $P_3 \sqsubseteq P_1$. However φ is not, and the model checker returns $\mathtt{false}$, with a path $s \to s'$ such that $s_{\downarrow V} = \bar{a}\bar{b}$ and $s'_{\downarrow V} = \bar{a}b$, which corresponds to a flip induced by P_1 but not by P_3. This provides the proof for $P_1 \not\sqsubseteq P_3$ and hence for $P_1 \not\equiv P_3$.

In summary, our approach is generic and can be used to verify equivalence and subsumption for any two sets of *ceteris paribus* statements, not necessarily expressed in the same preference language. Given two sets of preference statements P_1 and P_2, if there are $|V|$ nodes and $|E|$ edges in $\delta(P_1, P_2)$, then there are $O(|V|)$ states and $O(|E|)$ transitions in $K(P_1, P_2)$ by construction. Hence, the complexity of computing preference subsumption (and equivalence) is $(|V| + |E|) \times |\varphi|)$ as per the CTL model checking complexity [27]. Preliminary experiments indicate the feasibility of our approach for preferences expressed over up to 30 variables in a few seconds.

5   DISCUSSION

We have used examples of CP-nets [14] to explain necessary concepts and techniques for preference equivalence and preference subsumption checking; however, as in the case of Chapter 4, the approach presented in this chapter is applicable to all languages based on *ceteris paribus* semantics, i.e., those for which the semantics of the dominance relation can be given in terms of reachability within a graph of alternatives, including TCP-nets, CP-theories, and CI-nets. We conducted a preliminary feasibility study of our technique. The experiments indicate preferences expressed over up to 30 variables can be analyzed in less than a minute.

The method presented in this chapter can be extended to identify the set of all dominance relationships between alternatives in which the two sets of preferences differ, i.e., all possible ways one preference specification differs from another. In certain cases, this may be necessary to deduce the "degree" of differences between preferences of two or more agents. Finding these differences amounts to the following. First, we need to encode the proof of difference between preference specifications as obtained via model checking as a temporal logic property. This can be achieved easily as the proof of difference is a sequence of transitions that can be encoded using nesting of EX-formulas. In the second step, we relax the CTL formula used for finding equivalence by adding (disjunction) the proofs of differences that are already discovered. The verification of the relaxed CTL formula either returns true, i.e., the preference specifications are now equivalent modulo the already discovered proofs of differences; or returns false, revealing that other proofs of differences exist, which can be obtained by verifying the negation of the relaxed CTL formula as explained in Section 4.1.

CHAPTER 6

Ordering Alternatives With Respect to Preference

So far, we have seen examples of preference reasoning tasks that reason about dominance over pairs of alternatives or equivalence or subsumption of one preference specification with respect to another. In this chapter, consider the problem of ordering alternatives with respect to a given set of preferences. This problem finds applications in a variety of settings. For example, in recommendation systems, a user might want to consider the second or third best alternative when his most preferred alternative is unavailable for some reason; the amazon.com recommendation engine might want to present the user with the next best set of items in order of the user's preferences. Similarly, a policy maker might want to view the policies that are most preferred in order of preference. This chapter focuses on computing such an ordering for preferences stated using the *ceteris paribus* preference languages.

The problem of computing the next preferred alternative (ordering alternatives such that more preferred solutions are ahead of those that are less preferred) with respect to qualitative preferences has seen increased interest recently. Pilotto et. al. [71] introduced the problem of computing the next best solution with respect to a CP-net in the context of the stable marriage problem. Brafman et al. [19, 20] proved that finding the next-best solution in constraint satisfaction problems (CSPs) [89] is easy under some conditions, although it is hard in general for CSPs and weighted CSPs. Brafman et al. also show that the problem is easy in acyclic CP-nets and constrained acyclic CP-nets under some conditions; however finding the next-best solution with respect to general CP-nets (that can induce cyclic preferences among alternatives) is NP-hard. The corresponding complexity for more expressive preference languages such as TCP-nets, CI-nets, etc., have not yet been studied. Such algorithms are of importance in settings where suboptimal solutions may lead to undesirable consequences such as economic loss, customer dissatisfaction, and/or compromise of personal privacy. Specifically, the problem of finding the next best solutions respecting a set of preferences is also of central importance to evaluating skyline queries with preferences in the database setting [23, 24, 44, 55] and query answering for semantic search [63] using the ontological CP-net formalism [64].

1   OVERVIEW

In this chapter we show how to compute the next-preferred alternatives when the stated preferences can induce cyclic preferences between alternatives, which involves (a) automatically trans-

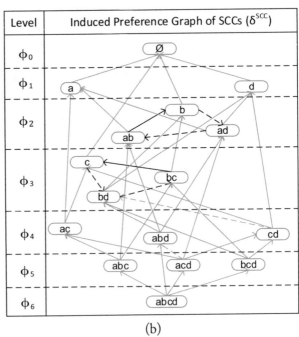

$p_1. \{d\}, \{\} : \{b\} \succ \{c\}$
$p_2. \{b\}, \{a\} : \{c\} \succ \{d\}$
$p_3. \{a\}, \{c\} : \{b\} \succ \{d\}$
$p_4. \{\}, \{a\} : \{b, d\} \succ \{c\}$
$p_5. \{\}, \{c\} : \{a, d\} \succ \{b\}$
$a :=$ Increase Logging Level
$b :=$ Restrict Access Privilege
$c :=$ Stop Service
$d :=$ Setup a Firewall

(a)   (b)

Figure 6.1: (a) Example CI-net preference specification, (b) induced preference graph $\delta(P)$.

lating a preference specification into the input model of NuSMV [26] model checker as we have seen earlier (Chapter 4) and (b) iteratively modifying and verifying the original preference model against a sequence of temporal properties (introduced in this chapter), whose counterexamples correspond to the desired ordering of next preferred solutions. There are two salient features of the approach developed in this chapter. First, we characterize the ordering computed here in terms of an extension of the partial order induced by the *ceteris paribus* semantics of the user preferences. Second, the approach is *applicable even when the user preferences are inconsistent*, i.e., may induce cycles in the preferences over solutions.

Example 6.1   Recall the example from Section 1.4 from Chapter 1 of ordering countermeasure sets starting from the most preferred one to the least that satisfy the security concerns of an administrator. The CI-net encoding of the preference statements for this example is given in Example 2.11 in Chapter 2.

In the rest of the chapter, we will use this as a running example. Figure 6.1(b) shows the corresponding induced preference graph. The dotted edges in the graph correspond to importance flips induced by the preference statements p_1 to p_5 and the solid edges correspond to monotonic-ity flips (see Chapter 2 Section 3.1). Note that there are two cycles—one is $ab \rightarrow b \rightarrow ad \rightarrow ab$; and another is $c \rightarrow bd \rightarrow bc \rightarrow c$, i.e., the preferences are inconsistent. Further, according to

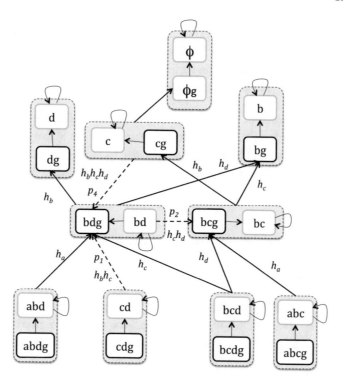

Figure 6.2: Kripke structure encoding for the semantics of example CI-net in Figure 6.1.

Definition 2.25 (see Chapter 2), b, ab, and ad are equivalent; similarly c, bd, and bc are equivalent.

We sometimes use "$P \models \gamma_1 R \gamma_2$" for $\gamma_1 R_P \gamma_2$ as a notational convenience. Edges of strongly connected components (SCCs) in Figure 6.1(b) resulting in equivalent alternatives are colored black (bold).

1.1   KRIPKE ENCODING

We will use ideas from Chapters 4 and 5 on constructing a Kripke structure model of the preference semantics of a preference specification. A part of the Kripke structure encoding constructed from the example CI-net in Figure 6.1 is shown in Figure 6.2. We recall some of the important details of the encoding using the running example of this chapter.

Example 6.2   Figure 6.2 presents a snapshot of the transitions to and from the states corresponding to the alternatives c, bc, and bd. The labels in the states denote the variables that evaluate to true. Pairs of states are grouped together in the figure to show that these pairs of states correspond to one alternative in the corresponding induced preference graph $\delta(P)$. The value of the global change variable g (introduced in the previous chapter for modeling) distinguishes the states in each pair: in one state of the pair the value of g is set to true and in the other it is set to false. Note that, in each pair, the value of the variables corresponding to preference variables in V are identical. We refer to the paired states as *meta-states*.

The transitions from both states in a meta-state is denoted by directed edges that start from the boundary of meta-state. The destination of these edges are states in some other meta-states where the value of g is true. These edges are labeled with the h-variables to denote the value of the preference variable (i.e., their presence or absence) that differ between the source and the destination states.

1.2   OBJECTIVE: COMPUTING AN ORDERED SEQUENCE

Computing an ordering of successively preferred alternatives with respect to a preference specification entails generating a sequence $\gamma_{01}, \gamma_{02}, \ldots, \gamma_{0m_0}, \ldots \gamma_{11}, \gamma_{12}, \ldots, \gamma_{1m_1}, \ldots, \gamma_{n1}, \ldots, \gamma_{nm_n}$ of alternatives where $\forall i \in [0, n] : \Phi_i = \cup_{j=1}^{m_i} \{\gamma_{ij}\}$, and alternatives in Φ_{i+1} are not preferred to any alternative in Φ_i. We refer set Φ_i as the set of *non-dominated* alternatives at level i – these alternatives are not dominated by any other alternative in Φ_{i+1}. For instance, as per Figure 6.1(b), $\Phi_0 = \{\emptyset\}$, $\Phi_1 = \{a, d\}$, and $\Phi_2 = \{b, ab, ad\}$.

A possible solution to the above problem will involve construction of $\delta(P)$ (Figure 6.1(b)) and then generation of a DAG over SCCs in $\delta(P)$. Construction of $\delta(P)$ will require considering all possible alternatives and identifying the edges between them by considering all possible pairs of alternatives in the context of the preference statements. Construction of a DAG over SCCs in $\delta(P)$ can be done in time linear to the number of nodes and edges in $\delta(P)$. Finally, this DAG needs to be explored in a way such that alternatives are output in order starting from level Φ_0 (top-level non-dominated set of alternatives). This can proceed by identifying the nodes in the DAG that do not have outgoing edges; the alternatives in these nodes belong to Φ_0. The DAG is then updated by removing these nodes; and the next set of nodes are identified which do not contain any outgoing edges—the alternatives in these nodes belong to Φ_1. The disadvantage with this approach to computing an ordered sequence of alternatives with respect to P is that the complexity of constructing $\delta(P)$ is exponential in the number of preference variables. Moreover, it may not be necessary to construct the entire $\delta(P)$ in order to generate an ordered sequence of alternatives, especially if only a small number of alternatives from the top of the ordering are needed. We next describe a model checking approach that generates an ordered sequence of alternatives that respects P in an anytime fashion, i.e., a sequence of k alternatives generated up to any given point represents the top-k alternatives with respect to P.

2 COMPUTATION OF ORDERED ALTERNATIVE SEQUENCE

The technique we will present involves automatic iterative updates to the Kripke structure as well as verification of carefully selected CTL temporal logic properties on these updates. In each iteration, we will obtain a set of alternatives that are most preferred among the alternatives that are yet to be output. In the following sections, we proceed by presenting certain properties of the induced preference graph (Section 2.1), which forms the basis of our iterative algorithm (Section 2.2).

2.1 DEALING WITH SCCS IN INDUCED PREFERENCE GRAPH

As noted in Section 1.2, our objective is to obtain an ordered sequence of alternatives. The sequence is further organized as a sequence of sets of outcomes $\Phi_0, \Phi_1, \ldots, \Phi_n$, where

- the alternatives in Φ_i are either equally preferred or are preferentially indifferent (see Definition 2.25), and

- the alternatives in Φ_{i+1} are not preferred to any alternative in Φ_i.

Our iterative algorithm will generate Φ_i at the i-th iteration. Φ_i is referred to as the *non-dominated* set at level i.

There are two types of alternatives that belong to the i-th non-dominated set Φ_i of alternatives.

Type I: alternatives in Φ_i, denoted Φ_i^I, that are not present in SCCs in the induced preference graph $\delta(P)$. $\Phi_i^I = \{\gamma \mid \gamma \in \Phi_i \land \forall \gamma' \in \Phi_i : (\gamma' \neq \gamma \Rightarrow \gamma' \not\asymp \gamma)\}$

Type II: alternatives in Φ_i, denoted Φ_i^{II}, that are indifferent to the Type I alternatives and are equivalent (equally preferred) to some other Type II alternative (i.e., belongs to an SCC in $\delta(P)$). $\Phi_i^{II} = \{\gamma \mid \gamma \in \Phi_i \land \exists \gamma' \in \Phi_i^{II} : \gamma' \asymp \gamma \land \forall \gamma" \in \Phi_i^I : (\gamma" \not\asymp \gamma)\}$

Note that there is at most one Type I set (Φ_i^I) and potentially multiple Type II sets ($\Phi_i^{II} = \{\Phi_i^{II_1}, \Phi_i^{II_2}, \ldots\}$) at each level i. In particular, there are as many sets of Type II sets of alternatives in level i as there are SCCs in $\delta(P)$ corresponding to that level, with $\Phi_i^{II_j}$ corresponding to the jth SCC at level i. Note that $\Phi_i = \Phi_i^I \cup (\cup_j \Phi_i^{II_j})$.

Example 6.3 For instance, in Figure 6.1(b),

- $\Phi_1^I = \{a, b\}$ and $|\Phi_1^{II}| = 0$ (absence of Type II sets)

- $\Phi_2^I = \{\}$ (Type I set is empty) and $\Phi_2^{II_1} = \{b, ab, ad\}$ and $|\Phi_2^{II}| = 1$ (presence of one Type II set).

We now state two characterizations of the Type I and Type II alternatives, which will be helpful to explain our algorithm for generating alternatives in order.

Proposition 6.4 *For the given preference specification P, $\forall i > 0$, $\forall \gamma \in \Phi_i^I$:*

a) $\exists \gamma' \in \Phi_{i-1} : \gamma' \succ \gamma$

b) $\nexists \gamma' \in \Phi_k, k > i : \gamma' \succ \gamma$

Proposition 6.4a states that each alternative in Φ_i^I has an edge in $\delta(P)$ (i.e., is strictly less preferred) to some alternative in Φ_{i-1}. Proposition 6.4b states that there is no alternative in any level greater than i to which there is an edge from any alternative in Φ_i^I. They state the necessary and sufficient conditions for alternatives of Type I to belong to a level i: a Type I alternative must be dominated by some alternative in level $i - 1$, and must not be dominated by any alternative in a level greater than i.

The conditions are similarly given for Type II sets of alternatives: at least one among a set of (equivalent) alternatives in Φ_i^{II} (that are involved in an SCC) must be dominated by some alternative in Φ_{i-1}; and none of the alternatives in the Type II set must be dominated by any alternative in a level greater than i. These are stated in Proposition 6.5.

Proposition 6.5 *For any preference specification P, $\forall i > 0 \; \forall j \; \forall \gamma \in \Phi_i^{II_j}$:*

a) $\exists \gamma' \in \Phi_i^{II_j}, \gamma'' \in \Phi_{i-1} : \gamma \bowtie \gamma' \wedge \gamma'' \succ \gamma'$

b) $\nexists \gamma' \in \Phi_k, k > i : \gamma' \succ \gamma$

The above propositions form the basis of our algorithm for computing an ordered sequence of alternatives $\gamma_{01}, \gamma_{02}, \dots, \gamma_{0m_0}, \dots \gamma_{11}, \gamma_{12}, \dots, \gamma_{1m_1}, \dots, \gamma_{n1}, \dots, \gamma_{nm_n}$ with respect to a preference specification P, such that $\forall i : \Phi_i = \cup_{j=1}^{m_i} \{\gamma_{ij}\}$.

At this point, before we delve into the algorithmic details, we state an assumption on the structure of the induced preference graph. This assumption allows us to present an algorithm that works for a restricted set of induced preference graphs. As subsequently pointed out, this assumption does not limit the generality of the algorithm itself.

Assumption 6.6 For any preference specification P, the following hold for Φ_0

a) The set Φ_0 always includes empty-set

b) The set Φ_0 is either a singleton set containing only $\emptyset$, i.e., $\Phi_0^I = \{\emptyset\}$ and $\Phi_0^{II} = \{\}$, or only contains Type II elements, i.e., $\Phi_0^I = \{\}$ and $\Phi_0^{II} \neq \{\}$

The above assumption states that the most preferred set of alternatives with respect to P always includes $\emptyset$. Note that this is trivially true for all CI-nets, where the induced preference graph is a lattice over the alternatives. In an acyclic CI-net there is exactly one set at the top of the lattice: either Ø or the entire set of alternatives. For other CI-nets, if Ø is involved in an SCC in $\delta(P)$, then all the alternatives in this SCC are equivalent to each other and to the alternative Ø; hence Φ_0 includes all alternatives in this SCC as Type II set. Otherwise, Φ_0 consists of just one alternative–Ø which is the unique, most preferred alternative with respect to P. For preference languages other than CI-nets, we could force the above assumption by introducing a dummy binary preference variable x_d which is also the most important of all variables with intra-variable preference $1 \succ 0$, which make the set of all alternatives with $x_d = 1$ as the non-dominated set at the top of the induced preference graph. Hence, the above assumption does not limit the generality of the algorithm presented in the following subsection.

2.2   ITERATIVE MODEL REFINEMENT AND PROPERTY RELAXATION

Algorithm 1 computes an ordering of alternatives by model checking specific CTL properties in a sequence and appropriately refining the Kripke structure model (encoding the preference semantics of a preference specification) and temporal properties with respect to the model. We proceed by first describing some important temporal properties that we will utilize to obtain the Type I and Type II elements at each level.

Property 6.7   Consider the objective of finding the most preferred alternative (a Type I element). The property to be verified is

$$\textbf{EX}g \tag{6.1}$$

EX describes a property for some (exists, **E**) path in the next state (**X**). That is, the property is satisfied in all states in $K(P)$ which can reach a state where g is set to true in one transition. The state that does not satisfy the property corresponds to the alternative which is most preferred because there is no transition from that state to another state where g is true—in other words, there is no transition from that state that corresponds to monotonicity or importance flips. The model checker will automatically find such a state as a counterexample witnessing the violation of the property in Equation 6.1.

Property 6.8   Consider an alternative γ that belongs to a Type II set. Therefore, there exist some alternatives that are equivalent to γ. The states corresponding to γ in Kripke structure $K(P)$ will, therefore, have some edges to states corresponding to another alternative. The objective is to find at least one equivalent alternative, say, γ'.

Recall that in $K(P)$, there are two states where the variables describing γ are true and in one of the two, the variable g is true. Let us denote these states as $\gamma_{\neg g}$ and γ_g. In $K(P)$, there is a transition from γ_g to $\gamma_{\neg g}$ and a transition from $\gamma_{\neg g}$ to $\gamma_{\neg g}$. The following temporal property is not satisfied by the states where γ holds and γ is a Type II alternative.

$$\gamma \Rightarrow \textbf{AX}\gamma \tag{6.2}$$

In the above, AX describes the property for all paths (A) in next (X)states. That is, the property states that from any state in $K(P)$ where γ holds, the state where γ holds can be visited along all paths via one transition. This is not true as γ is Type II alternative and there exists a path where in one step states satisfying γ can reach some other state not satisfying γ. The model checker can automatically find such a path leading to a state not satisfying γ as a counterexample proving the violation of the property in Equation 6.2. The last state in the resultant counterexample path corresponds to an alternative (e.g., γ') that is equivalent to γ.

Property 6.9   Consider the problem of verifying whether an alternative γ is a Type II alternative. For γ to be a Type II alternative, all alternatives reachable from γ should be able to reach γ in $\delta(P)$ (i.e., γ is part of cycle(s)); see Definition 2.25.

In the corresponding Kripke structure $K(P)$, this amounts to the fact that from all states reachable from the states corresponding to γ, the states corresponding to γ are reachable. This is captured by the following CTL property.

$$\gamma \Rightarrow \text{AGEF}\gamma \tag{6.3}$$

In the above, AG denotes the property in all states (G) along all paths (A). The EF denotes the property in some state in future (F) in some path (E). That is, the property in Equation 6.3 states that for any state s where γ holds ($\gamma \Rightarrow$), in all reachable states (say, t), there exists a path from t which eventually visits a state where γ holds as well. The property is satisfied by γ if and only if γ is a Type II alternative and there exists no other alternative strictly preferred to γ.

There are three major steps in our algorithm for iteratively computing the order sequence of alternatives from preference specifications.

Step 1. Computing Φ_0

Step 2. Computing Type I alternative at Φ_i for $i > 0$.

Step 3. Computing Type II alternatives at Φ_i for $i > 0$.

At each step, the Kripke structure is updated and/or appropriate CTL properties (see Properties 6.7, 6.8, and 6.9) are verified. In the following, we present our algorithm in terms of these steps.

Step 1: Computing Φ_0. As noted in the Assumption 6.6, $\emptyset \in \Phi_0$. It can be the only Type I alternative at Φ_0 or it can be part of the Type II elements. Hence, we start with $\Phi_0 = \{\emptyset\}$ and verify the CTL property $\text{SCC}(\Psi) := (\Psi \Rightarrow \text{AX } \Psi)$ in $K(P)$, where Ψ is initialized to Φ_0 (Line 3 in Algorithm 1 and Algorithm 2). This property (see Property 6.8) is satisfied in $K(P)$ if and only if $\emptyset$ is not included in an SCC in $\delta(P)$, i.e., $\emptyset$ is not a Type II alternative. This is the case in example Figure 6.1(b) ($\Phi_0 = \{\emptyset\}$).

On the other hand, if the property is not satisfied, the model checker returns a counterexample (a state in $K(P)$) corresponding to an alternative γ', that is in an SCC including $\emptyset$ in $\delta(P)$.

Algorithm 1 Computing the next-preferred alternatives

1: **procedure** NEXT-PREF
2:     $i = 0$
3:     $\Phi_i :=$ Find-Top-Cycle$(\emptyset, K_P)$
4:     $i = i + 1$
5:     Remove states corr. to alternatives in Φ_{i-1} from K_P
6:     **if** K_P is empty **then** return
7:     $\Phi_i = \{\}$
8:     Verify in K_P the property nds$(\Phi_i) : (\mathbf{EX}g) \vee \Phi_i$
9:     **if** counterexample (state corr. to γ) is generated **then**
10:         Add γ to Φ_i; Goto Line 8   ▷ **Type I alternative**
11:     **else**   ▷ **Check for Type II alternatives**
12:         $\Psi := \Phi_i$ ▷ **No more reconsidered as Type II**
13:         Include states in K_P corr. to alternatives in Φ_{i-1}
14:         Verify cand-SCC$(\Phi_{i-1}, \Psi) : (\neg\mathbf{EX}\ \Phi_{i-1}) \vee \Psi$ in K_P
15:         **if** counterexample corr. to γ is generated **then**
16:             **if** $\exists\gamma' \in \Psi : \gamma \supset \gamma'$ **then**   ▷ **Ignore alternatives in levels $> i$**
17:                 $\Psi := \Psi \cup \{\gamma\}$
18:                 Goto Line 14
19:             **end if**
20:             Remove states corresponding to the set Φ_{i-1} from K_P
21:             Verify verify-cand$(\gamma) := \gamma \Rightarrow \mathbf{AG}(\mathbf{EF}\gamma)$ in K_P
22:             **if** verify-cand(γ) holds **then**
23:                 $\Phi_i := \Phi_i \cup$ Find-Top-Cycle(γ, K_P) ▷ **Compute SCC**
24:                 $\Psi = \Psi \cup \Phi_i$ ▷ **Set up for next SCCs**
25:             **else**
26:                 Add γ to Ψ ▷ **No more reconsidered for Type II**
27:                 Goto Line 13
28:             **end if**
29:         **else** Goto Line 4   ▷ **No Type II in Φ_i; Φ_i computed**
30:     **end if**
31: **end procedure**

We add γ' to Ψ (see Line 5 in Algorithm 2). We proceed to identify the other alternatives that are present in this SCC. We iterate the model checking step using the property SCC(Ψ); note that the property is relaxed as Ψ now contains $\emptyset$ and γ'.

The model checking and addition of counterexample to Ψ (Lines 2–5 in Algorithm 2) are iterated until the property is satisfied. This step corresponds to Lines 2–3 in Algorithm 1. We next proceed to computing alternatives in $\Phi_1, \Phi_2, \ldots$ in order.

Step 2: Computing Φ_i^I ($i > 0$). We remove (Line 5 in Algorithm 1) the states in $K(P)$ corresponding to the alternatives present in Φ_{i-1} (using INVAR in the modeling language of NuSMV and add an invariant corresponding to the negation of alternatives in Φ_{i-1} in the modeling language of NuSMV).

Algorithm 2 Computing the Type II alternatives in an SCC

1: **procedure** FIND-TOP-CYCLE(γ, K_P)
2:    $\Psi := \{\gamma\}$
3:    Verify in K_P the property SCC(Ψ) : $\Psi \Rightarrow (\text{AX } \Psi)$
4:    **if** counterexample corr. to γ' is generated **then**
      $\triangleright$ **Type II alternative in a top SCC containing γ**
5:        Add γ' to Ψ; Goto Line 3
6:    **else** return Ψ
7: **end procedure**

We initialize a set Φ_i to empty set. We then verify to find non-dominating alternatives using the property $\text{nds}(\Phi_i) := \text{EX } g \vee \Phi_i$ in $K(P)$. Recall from Property 6.7, this property is satisfied in states corresponding to alternatives which are dominated by (less preferred to) some other alternative. If the property is not satisfied in $K(P)$, then the model checker returns a state as a counterexample. This state corresponds to some alternative that is not dominated by any other alternative at Φ_i. The alternative is added to Φ_i and the process is iterated to obtain all the Type I alternatives in Φ_i (see Lines 8–10 in Algorithm 1). Note that at each iteration, the property is relaxed by including the already obtained alternatives in Φ_i as a disjunct.

Following these step, in Figure 6.1(b), $\Phi_1^I = \{a, d\}$ is computed as Type I alternatives after $K(P)$ is updated by removing the states corresponding to alternatives in $\Phi_0 = \{\emptyset\}$.

Step 3: Computing alternatives in Φ_i^{II} ($i > 0$). Once Φ_i^I has been computed, we check for the existence of Type II alternatives, i.e., those that are involved in SCCs in $\delta(P)$. There may be multiple Type II alternative sets $\Phi_i^{II_j}$, each one resulting from a non-trivial SCC at level i.

Step 3a: Computing alternatives in $\Phi_i^{II_j}$ ($i > 0$). We identify alternatives that have one or more edges to some previous level alternatives in $\delta(P)$ (necessary condition for candidate alternatives in $\Phi_i^{II_j}$—see Proposition 6.5a).

We first include the states in $K(P)$ corresponding to alternatives in Φ_{i-1} that were removed after level $i - 1$ was computed. We verify the property cand-SCC(Φ_{i-1}, Ψ) : $\neg(\text{EX } \Phi_{i-1}) \vee \Psi$, where Ψ is the set of alternatives known **not to belong to** $\Phi_i^{II_j}$ (initially it is set to $\{\Phi_i\}$ to avoid considering Type I alternatives). The property is satisfied by states which correspond to alternatives in Ψ or which do not have any transition to states corresponding to alternatives in Φ_{i-1}. In the event the property is not satisfied, then the model checker obtains a state as a counterexample; such a state does not correspond to alternatives in Ψ and has a transition to a state corresponding to the alternative in Φ_{i-1} (i.e., dominated by or less preferred to the alternatives in Φ_{i-1}). In short, the counterexample generated by the model checker corresponds to an alternative γ that satisfies the necessary condition (Proposition 6.5a) for being a candidate for $\Phi_i^{II_j}$. If cand-SCC is satisfied, then there are no (more) Type II alternatives.

Let us denote the counterexample as γ. This candidate for Type II alternative γ must also satisfy the sufficient condition (Proposition 6.5b) before it can be included in $\Phi_i^{II_j}$. This can be verified by checking whether the alternative can also reach any alternative in level $> i$, in which case the sufficient condition is violated. There are two ways to check for this condition: is there an alternative preferred to γ as per (a) monotonicity or (b) importance flips.

First, if the alternative is a superset of any alternative that is ruled out from $\Phi_i^{\text{II}_j}$ (i.e., Ψ, see Line 12), then the alternative belongs to Φ_k with $k > i$ (according to monotonicity flip); see Lines 16–18.

Second, if the above check fails, then we proceed to remove the states corresponding to alternatives in Φ_{i-1} and verify the property `verify-cand`$(\gamma) : \gamma \Rightarrow \texttt{AGEF}\gamma$ in $K(P)$. Recall from Property 6.9, `verify-cand` is satisfied by states which belong to some SCC and cannot reach any state outside the SCC. That is, if the property is not satisfied by states corresponding to γ, then there exists some alternative in Φ_i that is preferred to γ as per importance flip and γ is ruled out from $\Phi_i^{\text{II}_j}$ (Lines 21, 25, 26).

Otherwise, γ belongs to $\Phi_i^{\text{II}_j}$ and we proceed to obtain the rest of the alternatives in $\Phi_i^{\text{II}_j}$ by invoking FIND-TOP-CYCLE in Algorithm 2 (Lines 21, 22).

Step 3b: Computing $\Phi_i^{\text{II}_j+1}$. After $\Phi_i^{\text{II}_j}$ is computed following Step 3a, we proceed to compute other SCCs (alternatives in $\Phi_i^{\text{II}_j+1}$) at the current level. Step 3a is reiterated by discarding the alternatives in $\Phi_i^{\text{II}_j}$ that have already been computed (Line 23 in Algorithm 1).

Once the alternatives Φ_i are obtained, we remove them from $K(P)$ by adding invariant corresponding to the negation of all the alternatives in Φ_i. Steps 2, 3 are repeated to compute alternatives in Φ_{i+1}. This step corresponds to Lines 27, 4–5 in Algorithm 1.

2.3   SAMPLE RUN OF THE ALGORITHM ON EXAMPLE IN FIGURE 6.1(B)

The run starts with Φ_0 containing $\emptyset$. At Line 3, `Find-Top-Cycle` is invoked to find any other alternatives that are equivalent to $\emptyset$. In our example, the property at Line 3 of `Find-Top-Cycle` is satisfied (see discussion in Property 6.8).

The run proceeds to Line 5 and the states corresponding to $\emptyset$ are removed from the Kripke structure (model refinement) and Φ_1 is initialized to empty set (Line 7). Next the Type I elements of Φ_1 are obtained from the counterexamples resulting from unsatisfiability of the property at Line 8 (see discussion in Property 6.7). All the Type I elements of Φ_1 are obtained in Lines 8–10; in our example, they are alternatives a and d.

Proceeding further, the next step is to obtain the first set $\Phi_1^{\text{II}_1}$ of Type II alternatives in Φ_1. The alternatives that cannot be in $\Phi_1^{\text{II}_1}$ are added to Ψ (Line 12), i.e., Ψ at this point is equal to $\{a, d\}$. The Φ_0 alternatives (the states corresponding to $\emptyset$) are brought back in the Kripke structure (Line 13). At Line 14, possible existence of Type II alternatives is checked using the property `cand-SCC`. In our example, the property is not satisfied and the model checker obtains a counterexample: a state which has a transition to some state corresponding to alternative $\emptyset$ in Φ_0 and which does not correspond to any alternative (a and d) in Ψ. Let such a state correspond to the alternative c (model checker can non-deterministically generate any valid counterexample).

Lines 16 check whether c is less preferred as per monotonicity to already obtained alternatives a and d in Ψ. The alternative c passes that test and the run proceeds to check whether c is less preferred to some other alternative that is not in Φ_0. To prepare for this, first states corresponding to alternative $\emptyset$ in Φ_0 is removed from the model (Line 19) and the property verify-cand is verified. The property is not satisfied by states corresponding to alternative c as there exists a path from such states to some other states (e.g., corresponding to alternative b) from where states corresponding to c is not reachable. The alternative c is, therefore, added to Ψ (Line 25) and the execution iterates from Line 13 (Line 26).

In this iteration, let the counterexample state generated as a result of unsatisfiability of property at Line 8 correspond to alternative b. As with alternative c, alternative b also passes the test that it is not dominated by the alternatives a and d via monotonicity. However, alternative b does not satisfy the property at Line 20 as the states corresponding to b can reach the states corresponding to alternative d and from there the states corresponding to b are not reachable. Therefore, b is discarded as well for Type II elements in Φ_1. The iteration continues from Line 13 again and this time the property at Line 14 is satisfied and the execution goes to Line 27. This marks the end of computing alternatives in Φ_1.

In the next iteration, starting from Line 4 alternatives in Φ_2 are computed. The first step is to find Type I alternatives of Φ_2. The property at Line 8 is satisfied as every existing state has a transition to some state with g set to true. Therefore, there is no Type I alternative in Φ_2. The run proceeds to Line 12 (Ψ being initialized to $\{\}$). The property at Line 14 is not satisfied and the model checker generates a counterexample: a state corresponding to some alternative that has an edge to alternatives a or d in Φ_1. Let the model checker obtain the state corresponding to the alternative b. The next steps (Lines 16–20) involve checking whether it is truly a Type II element in Φ_2. The alternative passes these tests, and alternatives involved in a cycle with b are generated by invoking Find-Top-SCC at Line 22 with γ equal to b. As a result, the alternatives ab and ad are obtained and added as Type II alternatives to Φ_2. The run moves to Line 13 again (from Line 26) to check for existence of other Type II alternatives. The check fails as the property at Line 14 is not satisfied and the execution proceeds to find alternatives in Φ_3.

The process terminates when all states in the Kripke structure have been removed and it is empty (Line 6).

2.4 NUMBER OF MODEL CHECKING CALLS

The number of calls for model checking can be obtained as follows. Let there be L SCCs, and the maximum size of the SCCs is M, then there are $O(L \times M)$ calls to verify the property in Line 3 in Algorithm 2. Next suppose there are N levels in the preferred ordering and the maximum size of Type I set at any level is $|\Phi^I|$, then the number of model checking calls to verify the property in Line 8 of Algorithm 1 is $O(N \times |\Phi^I|)$. Note that both these calls are inexpensive model checking calls as they verify one-step properties (involving X-operator).

Finally, if there are n alternatives, then the properties at Lines 14 and 20 can be potentially model checked against $O(n)$ alternatives. The worst case occurs when alternatives at level i have an edge to some alternative at level $i - 1$ and to some alternative at level $i - 2$. The model checker can potentially obtain counterexample alternatives from the lowest level after verifying the property at Line 14. Such alternatives will fail the checks for verifying the sufficient conditions, i.e., will fail the properties at Line 16 and Line 20. However, such special scenarios are likely to be infrequent and as a result the calls to verify property at Line 20 will be minimal. Note that in the absence of cycles, the algorithm can avoid Lines 11–26.

3 PROPERTIES OF NEXT-PREF

We prove that the sequence of sets $\Phi_0 \ldots \Phi_n$ corresponds to an *optimistic minimal weak order extension* of the partial order represented by $\delta(P)$ over the alternatives. This property guarantees that an alternative in the computed order is placed at the highest possible level in any weak order extending the partial order represented by (the edge relation of) $\delta(P)$. We use the following definition of a weak order in terms of a sequence of pairwise disjoint subsets of O, inspired by the characterization of weak orders by Bertet et al. [7].

Definition 6.10 Weak Order.   A binary relation $\mathcal{R} \subseteq O \times O$ is a weak order on O if and only if there exists a sequence $\Phi_0 \ldots \Phi_n$ of sets such that $\forall i < n : \Phi_i \subseteq O$ and the following hold: (a) $\forall i < j : \forall \gamma \in \Phi_i, \gamma' \in \Phi_j : \gamma' \mathcal{R} \gamma$; (b) $\forall i < j : \exists \gamma \in \Phi_i, \exists \gamma' \in \Phi_j : \gamma \mathcal{R} \gamma'$; and (c) $\forall \gamma, \gamma' \in \Phi_i : (\gamma \mathcal{R} \gamma' \wedge \gamma' \mathcal{R} \gamma) \vee (\gamma \mathcal{R} \gamma' \wedge \gamma' \mathcal{R} \gamma)$. ∎

In the above, condition (c) requires that if two elements in O are in Φ_i, they are either incomparable (first disjunct), or the relation $\mathcal{R}$ holds symmetrically for each other (second disjunct). This generalizes standard notions of strict and non-strict weak orders that allow only one of the conditions.

Definition 6.11 Optimiztic Minimal Weak Order Extension.   Let $\mathcal{R}, \mathcal{R}' \subseteq O \times O$ such that $\mathcal{R}$ is a partial order.

1. $\mathcal{R}'$ is a weak order extension of $\mathcal{R}$ if and only if it is a weak order and $\mathcal{R}' \supseteq \mathcal{R}$.

2. $\mathcal{R}'$ is a *minimal weak order extension* of $\mathcal{R}$ if and only if it is a weak order extension of $\mathcal{R}$, and there exists no weak order $\mathcal{R}''$ extending $\mathcal{R}$ such that $\mathcal{R}'' \subset \mathcal{R}'$. Equivalently,[1] $\mathcal{R}'$ corresponds to a sequence of subsets $\Phi_0 \ldots \Phi_n$ of O (Definition 6.10) s.t. $\forall i < n : \exists \gamma \in \Phi_i, \gamma' \in \Phi_{i+1} : \gamma \mathcal{R} \gamma'$.

3. $\mathcal{R}'$ is an *optimiztic minimal weak order extension* of $\mathcal{R}$ if and only if it is a minimal weak order extension, and it corresponds to a sequence $\Phi_0, \ldots, \Phi_n$ where there exists no weak order

---

[1] Minimal weak order extension was characterized by Bertet et al. [7].

extension $\mathcal{R}''$ which corresponds to a sequence $\Psi_0, \Psi_1, \ldots \Psi_m$ s.t. $\gamma \in \Phi_i \wedge \gamma \in \Psi_j \wedge j < i$.

∎

A minimal weak order extension of an order $\mathcal{R}$ is a weak order extension of $\mathcal{R}$ "closest" to $\mathcal{R}$. In an optimiztic minimal weak order extension of $\mathcal{R}$, an element is placed at the highest possible level in any weak order extending $\mathcal{R}$. Similarly, one can describe pessimistic minimal weak order extension of $\mathcal{R}$, where the elements are placed at the lowest possible level in any weak order extended $\mathcal{R}$.

Consider the following partial order over elements $\{A, B, P, D, E\}$:

$$A \succ E, \quad B \succ E, \quad P \bowtie D \tag{6.4}$$

There are multiple minimal weak order extensions; some of them are presented below.

```
Order 1.  Φ₀ = {A, B, P, D},  Φ₁ = {E}
Order 2.  Φ₀ = {A, B},        Φ₁ = {P, D, E}
Order 3.  Φ₀ = {A, P, D},     Φ₁ = {B, E}
```

Note that Order 1 is the optimiztic minimal weak order extension of the partial order; the elements are placed at the highest possible level. The Order 2, on the other hand, is one of the pessimistic extensions; one where the elements are placed in the lowest possible levels.

Example 6.12   For the running example in this chapter (see Figure 6.1(b)), note that the ordering is as follows.

$$\Phi_0 = \emptyset, \Phi_1 = \{a, d\}, \Phi_2 = \{b, ab, ad\}, \Phi_3 = \{c, bd, bc\}, \Phi_4 = \{ac, abd, cd\}, \Phi_5 = \{abc, acd, bcd\}, \Phi_6 = \{abcd\}$$

Consider an interesting case when the preference statements p_3 and p_5 are removed from the preference specification in Figure 6.1(a). Then the corresponding induced preferences (importance flips) $\{a, d\} \succ \{b\}$ and $\{a, b\} \succ \{a, d\}$ would be removed from the induced preference graph. In that case, the new optimiztic minimal weak order extension generated would be as follows.

$$\Phi_0 = \emptyset, \Phi_1 = \{a, b, d\}, \Phi_2 = \{ab, ad\}, \Phi_3 = \{c, bd, bc\}, \Phi_4 = \{ac, abd, cd\}, \Phi_5 = \{abc, acd, bcd\}, \Phi_6 = \{abcd\}$$

In the above, b is promoted to Φ_1 to maintain the optimiztic minimal weak order extension property, although the ordering would remain a weak order extension even if b is placed between Φ_1 and Φ_2. Note that in general, minimality is important to ensure that the generated extension of the partial order does not make any more distinction between alternatives than necessary. In other words, if minimality is not present, then the extension computed may introduce arbitrary levels in the weak order extension that are not warranted. An optimiztic minimal weak order extension ensures that the alternatives receive the maximum "ranking" possible in any given minimal weak order extension. This would be useful in settings such as the security example in this chapter, as well as in recommender systems where the user may want to obtain as many choices as possible that are most preferred at any given level.

Theorem 6.13 Correctness of Next-Pref (Algorithm 1).   *Given a preference specification P with preference relation $\succ$, NEXT-PREF generates the sequence of alternative sets $\Phi_0, \ldots, \Phi_n$, which corresponds to the optimiztic minimal weak order extension of the partial order induced by P with preference relation $\succ$.*

In this chapter, we focused on optimistic extension, as that intuitively provides users with the largest number of non-dominating possibilities as early as possible. However, if a certain application requires finding the pessimistic minimal weak order extension of a relation $\mathcal{R}$, our algorithm can do that as well. First, reverse the relation $\mathcal{R}$ to obtain $\mathcal{R}'$, i.e., $(A, B) \in \mathcal{R}$ results in $(B, A) \in R'$. Then, using our algorithm, compute the optimiztic minimal weak order extension of $\mathcal{R}'$, the reverse of which is the pessimistic minimal weak order extension of $\mathcal{R}$.

4   SUMMARY

We presented a generic model checking based technique to automatically compute a sequence of next preferred alternatives with respect to preference specifications. Our approach involves encoding the preference semantics directly as a Kripke model input of a model checker, along with a novel strategy to iteratively refine the Kripke model and to rewrite temporal properties to be verified such that the model checker automatically outputs a sequence of alternatives in order of preference. We also characterized the sequence as the optimiztic minimal weak order extension of the partial order induced by the preferences. This property guarantees that an alternative is placed at the highest possible level in any weak order respecting the preference semantics. To the best of our knowledge, this is the first approach to automatically compute a sequence of next preferred alternatives even when the induced preference graph has cycles.

CHAPTER 7

CRISNER: A Practically Efficient Reasoner for Qualitative Preferences

In this chapter, we present CRISNER (Conditional & Relative Importance Statement Network PrEference Reasoner),[1] a tool that provides exact and practically efficient reasoning about qualitative preferences in popular CP-languages.[2] The tool uses a model checking engine to translate preference specifications and queries into appropriate Kripke models and verifiable properties over them respectively, as per techniques outlined in the earlier chapters of this book. The salient features of CRISNER are: (1) exact and provably correct query answering for testing dominance, consistency with respect to a preference specification, and testing equivalence and subsumption of two sets of preferences; (2) automatic generation of proofs evidencing the correctness of answers produced by CRISNER to any of the above queries; (3) XML inputs and outputs that make it portable and pluggable into other applications. We also describe the extensible architecture of CRISNER, which can be extended to new preference formalisms based on *ceteris paribus* semantics that may be developed in the future.

1   OVERVIEW

Given a preference specification P, CRISNER first succinctly encodes the induced preference graph (IPG) $\delta(P)$ into a Kripke structure model K_P in the language of the NuSMV model checker (for details, please see Chapters 4 and 5). For reasoning with respect to each P, CRISNER generates the model K_P only once. Subsequently for each preference query q posed against P, CRISNER translates q into a temporal logic formula φ_q in computation-tree temporal logic (CTL) (see Chapter 4) such that $K_P \models \varphi_q$ if and only if q holds true according to the *ceteris paribus* semantics of P. CRISNER then queries the model checker with the model K_P and φ_q which either affirms q or returns false with a counterexample. For answering queries related to equivalence and subsumption checking of two sets of preferences P_1 and P_2, CRISNER constructs a combined IPG $K_{P_{12}}$ (see Chapter 5) and uses temporal queries in CTL to identify whether every dominance that holds in P_1 also holds in P_2 and vice-versa.

---

[1]CRISNER is available at http://www.ece.iastate.edu/~gsanthan/crisner.html. An earlier version of the tool, "iPref-R," is available at http://fmg.cs.iastate.edu/project-pages/preference-reasoner/.

[2]Henceforth, we will refer to the languages CP-nets, TCP-nets, and CP-theories collectively as "CP-languages" for brevity.

1.1   JUSTIFICATION OF QUERY ANSWERS

The answers to queries computed by CRISNER are exact and provably correct for dominance, consistency, equivalence, and subsumption queries. Because CRISNER uses the model checker for answering queries, CRISNER is also able to provide proofs or justifications to queries that returned `false`. CRISNER automatically builds proofs evidencing why the query did not hold true, by collecting and examining the model checker's counterexample and producing a sequence of preference statements whose application proves the correctness of CRISNER's result.

1.2   TOOL ARCHITECTURE

CRISNER is developed in pure Java and is *domain agnostic* in the sense that any set of variables with any domain can be included in a preference specification, although it is optimized for variables with binary domains. It accepts preference specifications and queries in an XML format, which provides a common and generic syntax using which users can specify preferences for CP-languages. The results (answers and proofs) for the corresponding queries are also saved in the form XML, so that the results can be further transformed into vocabulary that is more easily understandable by domain users. We describe the architecture of CRISNER and how it can be extended to other *ceteris paribus* preference formalisms that may be developed in the future.

CRISNER has been in development for over two years, and to our knowledge, CRISNER is one of the first attempts to develop practical tools for hard qualitative preference reasoning problems. We hope that CRISNER inspires the use of qualitative preference formalisms in practical, real world applications, and the development of further qualitative preference languages.

1.3   PREFERENCE QUERIES

Computing answers to preference queries with respect to a given preference specification in the *ceteris paribus* semantics amounts to making querying properties related to reachability on the induced preference graph. We consider the following preference queries that have been implemented in CRISNER in this chapter.

Given a preference specification P consisting of a set of preference statements $\{p_1, p_2 \ldots p_n\}$, and two outcomes $\alpha, \beta \in \mathcal{O}$, (a) **dominance testing** ($\alpha \succ_P \beta$) asks whether there is a sequence of improving flips from β to α in $\delta(P)$ (see Chapter 4) ; and (b) **consistency testing** asks whether the preferences induced by P are consistent, i.e., is there is a cycle in $\delta(P)$ (see Chapter 6). Given two preference specifications P_1, P_2 and two outcomes $\alpha, \beta \in \mathcal{O}$, (c) **preference subsumption** ($P_1 \sqsubseteq P_2$) asks whether $\alpha \succ_{P_1} \beta \Rightarrow \alpha \succ_{P_2} \beta$ (see Chapter 5); and (d) **preference equivalence** ($P_1 \equiv P_2$) asks whether $P_1 \sqsubseteq P_2$ and $P_2 \sqsubseteq P_1$ (see Chapter 5).

The technique for encoding P computing the answers to the various preference queries has been covered in the above mentioned chapters respectively. The interested reader will find it useful to review and refer to these chapters for details on the modeling techniques and the formulation of the corresponding temporal logic formulas.

2   XML INPUT LANGUAGE

CRISNER accepts a preference specification for any of the CP-languages in an XML format. The preference specification consists of a declaration of the preference variables, their domains, and a set of preference statements. Each preference statement is of the form discussed in Chapter 2, and expresses an intra-variable and/or relative importance preference relation over the domain of a variable.

2.1   DEFINING PREFERENCE VARIABLES

Figure 7.1 shows part of a preference specification defining variables and their domains. The preference variable a has a binary domain with values 0 and 1, whereas x has a domain $\{0, 1, 2\}$. Note that CRISNER supports domain valuations with string values that are combinations of letters and numbers, as allowed by NuSMV.

```
<VARIABLE>
  <NAME>a</NAME>
  <DOMAIN-VALUE>0</DOMAIN-VALUE>
  <DOMAIN-VALUE>1</DOMAIN-VALUE>
</VARIABLE>
```
(a)

```
<VARIABLE>
  <NAME>x</NAME>
  <DOMAIN-VALUE>0</DOMAIN-VALUE>
  <DOMAIN-VALUE>1</DOMAIN-VALUE>
  <DOMAIN-VALUE>2</DOMAIN-VALUE>
</VARIABLE>
```
(b)

Figure 7.1: XML encoding of definitions of preference variable a with domain size 2 and 3.

2.2   SPECIFYING CONDITIONAL PREFERENCE STATEMENTS

The listing in Figure 7.2 shows a portion of a preference specification that declares preferences over values of the variable c conditioned on the variables b and a respectively. The STATEMENT-ID is an identifier to a unique preference statement in a preference specification. The VARIABLE tag identifies the variable on whose domain preferences are being specified. The CONDITION tag is used to specify the condition for the preference statement as an assignment of the parent variable to its domain value. Note that there can be multiple conditions, in which case there will be multiple CONDITION tags, or no conditions (unconditional preference) as well in a preference statement. In addition, there can also be multiple preferences for a variable, e.g., if there is a variable with domain of $0, 1, 2$ then to specify the total order $0 \succ 1 \succ 2$ one would encode $0 \succ 1$ as one preference followed by $1 \succ 2$. CRISNER requires that the variable names and their assignments match with the preference variable declarations in the file; otherwise the tool reports an appropriate error stating that the variable is not defined in the preference specification.

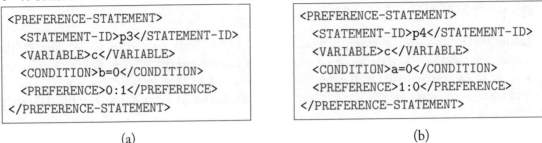

```
<PREFERENCE-STATEMENT>
  <STATEMENT-ID>p3</STATEMENT-ID>
  <VARIABLE>c</VARIABLE>
  <CONDITION>b=0</CONDITION>
  <PREFERENCE>0:1</PREFERENCE>
</PREFERENCE-STATEMENT>
```

(a)

```
<PREFERENCE-STATEMENT>
  <STATEMENT-ID>p4</STATEMENT-ID>
  <VARIABLE>c</VARIABLE>
  <CONDITION>a=0</CONDITION>
  <PREFERENCE>1:0</PREFERENCE>
</PREFERENCE-STATEMENT>
```

(b)

Figure 7.2: XML encoding of conditional preference statements p_3 and p_4 in a CP-net.

2.3   SPECIFYING RELATIVE IMPORTANCE PREFERENCES

In order to allow specification of relative importance of one variable over another, as in a TCP-net, CRISNER allows the tag REGARDLESS-OF within a preference statement. Figure 7.3(a) declares a preference statement that says (in addition to the fact that $a = 0 \succ_a a = 1$) that a is relatively more important than b. In order to specify relative importance of one variable over a set of other variables (simultaneously) as allowed by a CP-theory, the user can specify multiple REGARDLESS-OF tags within the same preference statement. For instance, Figure 7.3(b) shows a preference statement that declares that a is relatively more important than $\{b, c\}$.

```
<PREFERENCE-STATEMENT>
  <STATEMENT-ID>p1</STATEMENT-ID>
  <VARIABLE>a</VARIABLE>
  <PREFERENCE>0:1</PREFERENCE>
  <REGARDLESS-OF>b</REGARDLESS-OF>
</PREFERENCE-STATEMENT>
```

(a)

```
<PREFERENCE-STATEMENT>
  <STATEMENT-ID>p1</STATEMENT-ID>
  <VARIABLE>a</VARIABLE>
  <PREFERENCE>0:1</PREFERENCE>
  <REGARDLESS-OF>b</REGARDLESS-OF>
  <REGARDLESS-OF>c</REGARDLESS-OF>
</PREFERENCE-STATEMENT>
```

(b)

Figure 7.3: XML encoding of conditional preference statements p_3 and p_4 in a CP-net.

The sample XML input file listings for the preference specifications used for the cyberdefense example from Chapter 1 are provided in Appendix B.

3   ENCODING PREFERENCES AS SMV MODELS

CRISNER encodes Kripke models for a preference specification as described earlier in Chapters 4, 5, and 6 for the corresponding preference reasoning tasks. Here we discuss specific constructs in the NuSMV model checker used by CRISNER for this encoding using the examples introduced in this chapter.

3.1   ENCODING PREFERENCE VARIABLES & AUXILIARY VARIABLES

In order to encode the CP-net P_1 in our earlier example, CRISNER generates the code for the SMV model as shown in Figure 7.4. We explain the translation of a preference specification into an SMV model by CRISNER through this example.

CRISNER defines just the main module, with three variables corresponding to the preference variables in P_1 and another variable g, which we will explain shortly. We overload a, b, c to refer to variables in the Kripke model and variables in the preference specification, hence valuations of a, b, c in a state s in the Kripke model respectively correspond to the valuations of the preference variables a, b, c in the preference specification P. The variables a, b, c are *state* variables in the SMV model, i.e., their valuations stored by the model checker for each state explored during model checking.

The IVAR variables cha, chb, chc are modeled as *input* variables, i.e., their valuations are not stored as part of each state. The model checker initializes them non-deterministically for each state so that all paths are open for exploration by the model checker during verification. Each preference statement is translated into an appropriate guard condition for a transition in the Kripke model, and the semantics of variables cha, chb, chc either allows or disallows the change in the value of the corresponding preference variable a, b, c, in accordance with the improving flip semantics.

Identifying transitions corresponding to improving flips. The additional g variable is defined to be *true* exactly when the model checker transitions from a state corresponding to one outcome to a state corresponding to another outcome (not transitions between states corresponding to the same outcome). Hence, we can conveniently refer to transitions in the Kripke models that correspond to improving flips by constraining g to have valuation 1 in the destination state.

Referencing start states explored by NuSMV. The FROZEN variables a_0, b_0, c_0 are constrained to be fixed with the values of the variables a, b, c respectively at the start of the model checking algorithm via the DEFINE and INIT constructs. This allows us to refer to the state non-deterministically picked by the model checker as the start of model exploration using $start$. This is used for computing consistency, preference subsumption, and preference equivalence.

3.2   ENCODING PREFERENCE STATEMENTS

Encoding Intra-variable Preferences. To encode an intra-variable preference statement for a variable x with a condition ρ on the other variables, the next(x) construct encodes guards such that the valuations of the other variables correspond to those in the condition ρ, and valuation of chx is 1 while all other ch variables are set to 0. As an example, next(a) includes a transition such that c changes from 1 to 0 precisely when $b = 0$ and $chc = 1$ ($cha = 0, chb = 0$, allowing only c to change in that transition), which corresponds to the improving flip induced by p_3 conditioned on the value of a in the CP-net P_1 (Figure 7.4).

```
MODULE main                          ASSIGN
VAR                                    next(a) := case
  a : {0,1};                             a=1 & cha=1 & chb=0 & chc=0 : 0;
  c : {0,1};                             TRUE : a;
  b : {0,1};                           esac;
  g : {0,1};                           next(c) := case
FROZENVAR                                c=1 & b=0 & cha=0 & chb=0 & chc=1 : 0;
  a_0 : {0,1};                           TRUE : c;
  b_0 : {0,1};                         esac;
  c_0 : {0,1};                         next(b) := case
IVAR                                      b=0 & c=0 & cha=0 & chb=1 & chc=0 : 1;
  cha : {0,1};                           TRUE : b;
  chb : {0,1};                         esac;
  chc : {0,1};                         next(g) := case
DEFINE                                    a=1 & cha=1 & chb=0 & chc=0 : 1;
start := a=a_0                            c=1 & b=0 & cha=0 & chb=0 & chc=1 : 1;
     & b=b_0 & c=c_0;                     b=0 & c=0 & cha=0 & chb=1 & chc=0 : 1;
                                          TRUE: 0;
INIT start=TRUE;                        esac;
```

Figure 7.4: SMV code for Kripke model encoding $\delta(P_1)$.

Encoding Relative Importance. For modeling relative importance preference statements, multiple IVAR variables can be assigned 1 in guard conditions such that the more important and less important preference variables can change in the same transition—corresponding to an improving flip for relative importance. For example, Figure 7.5 shows a snippet of the SMV code that models the transitions arising from the relative importance of a over b as in the TCP-net P_3 shown in Figure 2.1(a)(c). Note that cha and chb are set to 1 for the second guard condition of next(b), allowing a to change to a preferred value trading off b. In order to model relative importance as in a CP-theory where one variable is more important than multiple others, a similar encoding is used, except that all the corresponding ch variables are set to 1.

```
next(b) :=  case
     b=0 & c=0 & cha=0 & chb=1 & chc=0 : 1;
     a=1 & cha=1 & chb=1 & chc=0 : {0,1};
     TRUE : b;
  esac;
```

Figure 7.5: Encoding relative importance preferences for the TCP-net P_3.

3.3   JUSTIFICATION OF QUERY RESULTS

In addition to answering preference queries posed against preference specifications, CRISNER also provides a justification of the result where appropriate. In order to obtain justification, CRISNER uses the counterexamples returned by the NuSMV model checker whenever a temporal logic formula is not satisfied.

Extracting a Proof of Dominance. In the case of a dominance query, if CRISNER returns true, we construct another temporal logic formula that states the negation of the dominance relationship, which obtains a sequence of outcomes corresponding to an improving flipping sequence from the lesser preferred to the more preferred outcome from the model checker. Suppose that we want to obtain proof that an alternative α dominates another alternative β. This means that $\varphi_\beta \rightarrow EF\varphi_\alpha$ holds. We then verify $\neg(\varphi_\beta \rightarrow EF\varphi_\alpha)$, which obtains a sequence of states in the Kripke model corresponding to an improving flipping sequence from β to α from the model checker corresponding to an improving flipping sequence from β to α which serves as the proof of dominance.

Extracting a Proof of Inconsistency. In the case of a consistency query (see Section 1.3) CRISNER returns a sequence of outcomes corresponding to an improving flipping sequence from an outcome to itself (indicating a cycle in the induced preference graph) whenever the preference specification input is inconsistent.

Extracting a Proof of Non-subsumption. For a preference subsumption query $P_1 \sqsubseteq P_2$, CRISNER provides an improving flip from one outcome to another induced by P_1 but not induced by P_2 whenever the query does not hold.

In the above, counterexamples returned by NuSMV are in terms of states and transitions in the Kripke model; CRISNER parses and transforms the counterexamples back into a form that relates to the preference variables, outcomes, and improving flips in the induced preference graph of the preference specification, and saves it in an XML format.

4   ARCHITECTURE

CRISNER is built using the Java programming language.[3] The architecture of CRISNER consists of several components as depicted in Figure 7.8.

The XML parser is used to parse the preference specifications and preference queries input by the user.[4] The CP-language translator is a critical component that constructs the SMV code for the Kripke model for the preference specification input. It declares the necessary variables with their domains, sets up the DEFINE, TRANS, and INIT constraints, and finally generates guard conditions for enabling transitions corresponding to improving flips induced by each preference statement (as discussed in Chapter 4).

---

[3]Third party libraries used by CRISNER are listed in the project site [80].
[4]While currently CRISNER does not use XML schema or DTD to validate the XML input, we plan to enforce that in the future.

CRISNER provides two interfaces for preference reasoning. The first is a simple command line menu-driven console interface where the user can provide either (a) one preference specification as input and then use the menu to pose dominance and consistency queries, or (b) two preference specifications as input and pose a subsumption or equivalence query. The answer (true/false) obtained and the justification for the answer (where possible) is provided on the console. Figure 7.6 shows the command line interface in which the user queries for consistency and obtains a negative result along with a cycle in the induced preference graph as evidence. Figure 7.7 shows a similar interaction of a user with CRISNER for preference equivalence testing with respect to two preference specifications. Another way of using CRISNER is to specify preference queries in an XML file that contains a dominance or consistency or preference equivalence or subsumption query, and identifies the preference specification against which the query should be executed. The query translator component parses queries specified in XML format and provides it to the Reasoner component.

```
C:\> java -jar CRISNER-full.jar -s samples\cycle-cpnet.xml -m nusmv

Parsing preference specification ... samples\cycle-cpnet.xml

Reasoning options:
[1] Test Dominance
[2] Test Dominance Performance
[3] Test Consistency
[4] Test Subsumption*
[5] Test Equivalence*
[9] Exit
(* - For subsumption and equivalence, we currently require both specifications to have
the same set of preference variables and respective domains.)
Enter option: 3
Cycle found.
Sequence: [c = 1, a = 0, b = 0] -> [c = 0, a = 0, b = 0] -> [c = 1, a = 0, b = 0]

Result: FALSE; Proof: Sequence: [c = 1, a = 0, b = 0] -> [c = 0, a = 0, b = 0] -> [c =
    1, a = 0, b = 0]
```

Figure 7.6: User interaction via command line interface to CRISNER to verify the consistency of a preference specification.

Further details about the XML tags and examples of preference specification, preference queries, and sample SMV code generated by CRISNER are available from CRISNER's project website http://www.ece.iastate.edu/~gsanthan/crisner.html. The same website also contains a guide to navigating and using the menu-driven console user interface to CRISNER.

```
C:>java -jar CRISNER-full.jar -s samples\nocycle-cpnet.xml -m nusmv

Parsing preference specification ... samples\nocycle-cpnet.xml

Reasoning options:
[1] Test Dominance
[2] Test Dominance Performance
[3] Test Consistency
[4] Test Subsumption*
[5] Test Equivalence*
[9] Exit
(* - For subsumption and equivalence, we currently require both specifications to have
    the same set of preference variables and respective domains.)
Enter option: 5
Please enter another XML preference specification to test equivalence:
Enter the location of the XML file: samples\nocycle-tcpnet.xml
Is nocycle-cpnet.xml subsumed by nocycle-tcpnet.xml ?
Subsumption holds.
Is nocycle-tcpnet.xml subsumed by nocycle-cpnet.xml ?
Subsumption does not hold.
false; nocycle-cpnet.xml is not equivalent to nocycle-tcpnet.xml
```

Figure 7.7: User interaction via command line interface to CRISNER to verify preference equivalence.

The Reasoner is another critical component in CRISNER that constructs a temporal logic formula corresponding to the preference query posed by the user, and invokes the NuSMV model checker to verify the formula. The result and any counterexamples generated by the model checker are parsed by the Results Translator, and saved in XML format by the XML Encoder. If a counterexample is applicable to the preference query, then the Justifier parses the XML output and executes any followup queries on the model checker (e.g., verification of the negation of the dominance query) to provide the user with the appropriate proof.

4.1   EXTENDING CRISNER

CRISNER currently supports the CP-net, TCP-net, and CP-theory formalisms and can perform dominance, consistency, preference subsumption, and preference equivalence reasoning for preference specifications in these languages. Some possible ways to extend the tool to support other preference formalisms are given below.

Supporting Other Preference Languages. CRISNER can be extended to support another qualitative preference language, as long as the semantics of the language is described in terms of an induced preference graph. In particular, the XML parser must be extended to support the syntax

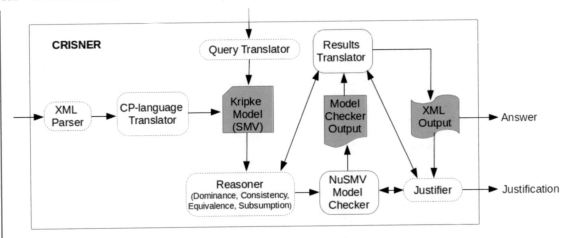

Figure 7.8: Architecture and components of CRISNER preference reasoner.

of the new language; and the CP-language translator must be extended to generate SMV code according to the semantics of the new language. Support for the conditional importance network or CI-net language (see Chapter 2, Section 2.6) can be added by implementing the encoding of the CI-net semantics according to Section 2.1 of Chapter 4. One can think of supporting a new preference formalism that allows expression of preferences of one partial assignment of variables over another. In such a case, the CP-translator component can be extended to include a translation for such a preference statement into guard conditions in the Kripke model representing the induced preference graph.

Supporting Other Preference Queries. CRISNER can also be extended to support new preference queries, for example, computing a weak order or total order extension of the partial order induced by a preference specification. In this case the Query Translator and Reasoner should be extended to translate the new queries into one or more appropriate temporal logic formulas (as described in [69]) and the Justifier should be extended to construct and execute followup queries that obtain proofs for the answers. The interested reader may try to implement the Next-Pref algorithm presented in Chapter 6 for a better understanding of the model checking approach to preference reasoning.

Adding Support for Other Model Checkers. CRISNER currently uses the NuSMV model checker version 2.5.4 for verifying temporal queries specified in CTL. However, one may use any other model checker which supports CTL in place of NuSMV, such as Cadence SMV [61]. The required change would then be to modify (a) the CP-translator module to generate Kripke structure models that are compatible with the input language of the new model checker, and (b) the Results Translator to parse the traces of the new model checker and obtain results of the specified queries.

Adding GUI support. CRISNER currently uses the console interface for user interaction, such as specifying preference queries and reporting their results. Because CRISNER uses portable XML formats for specifying inputs and reporting outputs of preference reasoning tasks, one may develop a graphical user interface (GUI) front-end for CRISNER that binds to these XML formats. Integrating such a GUI would be straightforward if the GUI is written to support the XML formats of CRISNER.

4.2   SCALABILITY

While we have not yet performed a systematic experiment studying CRISNER's runtime performance for preference specifications of different sizes (number of preference variables), our preliminary tests have revealed that CRISNER answers dominance (including the computation of justification when applicable) in less than a minute on average for up to 30 variables on a 8GB Corei7 Windows 7 desktop. Although CRISNER allows variables with domain size $n > 2$, the model checking performance degrades quickly with increasing n; this can be alleviated by configuring the NuSMV model checker to use multi-way decision diagrams [4].

5   CONCLUDING REMARKS

We presented CRISNER, a tool for specifying and reasoning with qualitative preference languages such as CP-net, TCP-net and CP-theory. CRISNER translates preference specifications and queries with respect to those provided in XML format into a Kripke structure and corresponding temporal logic (CTL) queries in the input language of the NuSMV model checker. Currently CRISNER supports dominance, consistency, preference equivalence, and subsumption testing for the above languages. The obtained results from the model checker, including proofs of dominance, inconsistency, or non-subsumption, are translated by CRISNER back in terms of the vocabulary of the input preference specification and saved in XML format. CRISNER's architecture supports extension to other preference queries and preference languages such as CI-nets whose semantics are in terms of the induced preference graph.

CHAPTER 8

Postscript

The ability to represent and reason about preferences over a set of alternatives is central to rational decision making. Preferences have been the subject of study in many disciplines including decision theory [38, 54], social choice [78], game theory [67], and other subfields of economics [62]. Artificial intelligence (AI) brings a fresh perspective to the study of preferences, and studies in AI are concerned with how to efficiently represent and reason about preferences especially when the preferences involve multiple attributes, resulting in a large combinatorial space of alternatives.

In this book, we have provided a tutorial introduction to modern techniques for reasoning about qualitative preferences with respect to a set of alternatives. We focused on CP-net, TCP-net, CP-theory, and CI-net formalisms, and discussed their semantics according to the *ceteris paribus* interpretation of preferences. Given the hardness of most reasoning tasks such as dominance testing, preference equivalence, and subsumption testing, and ordering outcomes with respect to a given set of preferences, we described an approach to reasoning with qualitative preferences by translating them into a temporal logic model checking problem.

Our treatment has focused on reasoning about preferences of a single agent. In many settings however, the preferences of multiple agents [32, 91] have to be considered in the decision making process. This presents additional challenges: the preferences of different agents may conflict with each other. For example, while Jane wants a low-rent apartment in midtown Manhattan, Jane's husband John prefers a single family home on Long Island. One outcome could be preferred to another based on one agent's preferences, while the opposite may be inferred from the other agent's preferences. Studies in the field of social choice consider conflict resolution strategies [72, 78] that minimally compromise individual preferences agents, or take advantage of the organizational hierarchies that force the preferences of some agents to take precedence over the preferences of others.

Our treatment has focused on preferences expressed directly over the alternatives and their attributes. However, in many applications, the alternatives additionally have a compositional structure: each alternative is composed of a collection of components that work together in order to achieve a desired functionality, e.g., web services [81], team members [57], software systems [68], buildings [83], plans [51], etc. A student wanting to design his program of study by choosing a set of courses from the course catalog (see Section 1.2 from Chapter 1) also illustrates compositional structure of alternatives. Recent work [84] has led to techniques for reasoning about preferences over compositional systems by considering the preferences over attributes of alternatives in terms of the preference over the attributes of their constituent components.

Further complications in reasoning about preferences arise from uncertainty associated with the preferences or choices. Extending the methods described in this book to settings with uncertainty remains an interesting area for further research.

APPENDIX A

SMV Model Listings

Here we provide listings of the SMV models generated by the preference reasoning tool CRISNER (see Chapter 7) for the preference specifications P^{CP}, P^{TCP}, and P^{CPT} introduced in Chapter 1 first, and used in Chapters 2 and 4. Here you will also find traces of output produced by model checker NuSMV for specific CTL queries executed on these models. The reader will find this a useful reference especially when studying Chapters 4 and 7.

1   SMV MODEL LISTING FOR P^{CP}

```
MODULE main

VAR
  f : {Fix-Available,Fix-Unavailable};
  a : {High,Low};
  e : {Code-Available,Code-Unavailable};

IVAR
  cha : {0,1};
  che : {0,1};
  chf : {0,1};

ASSIGN
  next(f) :=
    case
      f=Fix-Available & e=Code-Available & cha=0 & che=0 & chf=1 : Fix-Unavailable;
        -- #p3 : [e=Code-Available] => f=[Fix-Unavailable:Fix-Available]  >> [].
      TRUE : f;
    esac;
  next(a) :=
    case
      a=High & cha=1 & che=0 & chf=0 : Low;
        -- #p2 : null => a=[Low:High]  >> [].
      TRUE : a;
    esac;
  next(e) :=
    case
      e=Code-Unavailable & cha=0 & che=1 & chf=0 : Code-Available;
        -- #p1 : null => e=[Code-Available:Code-Unavailable]  >> [].
      TRUE : e;
    esac;
```

2   DOMINANCE QUERY AND NUSMV OUTPUT FOR P^{CP}

```
SPEC (a=High & e=Code-Available & f=Fix-Available ->
            !EX EF (a=Low & e=Code-Available & f=Fix-Unavailable))

-- specification (((a = High & e = Code-Available) & f = Fix-Available)
-> !(EX (EF ((a = Low & e = Code-Available) & f = Fix-Unavailable)))) is false.
-- as demonstrated by the following execution sequence.
Trace Description: CTL Counterexample
Trace Type: Counterexample
-> State: 1.1 <-
  f = Fix-Available
  a = High
  e = Code-Available
-> Input: 1.2 <-
  cha = 0
  che = 0
  chf = 1
-> State: 1.2 <-
  f = Fix-Unavailable
-> Input: 1.3 <-
  cha = 1
  chf = 0
-> State: 1.3 <-
  a = Low
```

3   SMV MODEL LISTING FOR P^{TCP}

```
MODULE main

VAR
  a : {High,Low};
  e : {Code-Available,Code-Unavailable};
  f : {Fix-Available,Fix-Unavailable};

IVAR
  cha : {0,1};
  che : {0,1};
  chf : {0,1};

ASSIGN
  next(a) :=
    case
      a=High & cha=1 & che=1 & chf=0 : Low;
        -- #p'2 : [] => a=[Low:High] >> [e].
      TRUE : a;
```

```
      esac;
  next(e) :=
      case
        a=High & cha=1 & che=1 & chf=0 : {Code-Available,Code-Unavailable};
          -- #p'2 : [] => a=[Low:High]  >> [e].
        e=Code-Unavailable & cha=0 & che=1 & chf=0 : Code-Available;
          -- #p1 : null => e=[Code-Available:Code-Unavailable]  >> [].
        TRUE : e;
      esac;
  next(f) :=
      case
        f=Fix-Available & e=Code-Available & cha=0 & che=0 & chf=1 : Fix-Unavailable;
          -- #p3 : [e=Code-Available] => f=[Fix-Unavailable:Fix-Available]  >> [].
        TRUE : f;
      esac;
```

4   DOMINANCE QUERY AND NUSMV OUTPUT FOR P^{TCP}

```
SPEC (a=High & e=Code-Available & f=Fix-Available ->
            !EX EF (a=Low & e=Code-Unavailable & f=Fix-Unavailable))

-- specification (((a = High & e = Code-Available) & f = Fix-Available)
-> !(EX (EF ((a = Low & e = Code-Unavailable) & f = Fix-Unavailable))))  is false
-- as demonstrated by the following execution sequence
Trace Description: CTL Counterexample
Trace Type: Counterexample
-> State: 1.1 <-
  a = High
  e = Code-Available
  f = Fix-Available
-> Input: 1.2 <-
  cha = 1
  che = 0
  chf = 0
-> State: 1.2 <-
-> Input: 1.3 <-
  cha = 0
  chf = 1
-> State: 1.3 <-
  f = Fix-Unavailable
-> Input: 1.4 <-
  cha = 1
  che = 1
  chf = 0
-> State: 1.4 <-
  a = Low
  e = Code-Unavailable
```

5   SMV MODEL LISTING FOR P^{CPT}

```
MODULE main

VAR
  f : {Fix-Available,Fix-Unavailable};
  a : {High,Low};
  e : {Code-Available,Code-Unavailable};

IVAR
  cha : {0,1};
  che : {0,1};
  chf : {0,1};

ASSIGN
  next(f) :=
    case
      a=High & cha=1 & che=1 & chf=1 : {Fix-Available,Fix-Unavailable};
        -- #p''2 : [] => a=[Low:High]  >> [e, f].
      f=Fix-Available & e=Code-Available & cha=0 & che=0 & chf=1 : Fix-Unavailable;
        -- #p3 : [e=Code-Available] => f=[Fix-Unavailable:Fix-Available]  >> [].
      TRUE : f;
    esac;
  next(a) :=
    case
      a=High & cha=1 & che=1 & chf=1 : Low;
        -- #p''2 : [] => a=[Low:High]  >> [e, f].
      TRUE : a;
    esac;
  next(e) :=
    case
      e=Code-Unavailable & cha=0 & che=1 & chf=0 : Code-Available;
        -- #p1 : null => e=[Code-Available:Code-Unavailable]  >> [].
      a=High & cha=1 & che=1 & chf=1 : {Code-Available,Code-Unavailable};
        -- #p''2 : [] => a=[Low:High]  >> [e, f].
      TRUE : e;
    esac;
```

6   DOMINANCE QUERY AND NUSMV OUTPUT FOR P^{CPT}

```
SPEC (a=High & e=Code-Unavailable & f=Fix-Unavailable ->
        !EX EF (a=Low & e=Code-Available & f=Fix-Available))

-- specification (((a = High & e = Code-Unavailable) & f = Fix-Unavailable)
-> !(EX (EF ((a = Low & e = Code-Available) & f = Fix-Available))))  is false.
-- as demonstrated by the following execution sequence.
```

```
Trace Description: CTL Counterexample
Trace Type: Counterexample
-> State: 1.1 <-
  f = Fix-Unavailable
  a = High
  e = Code-Unavailable
-> Input: 1.2 <-
  cha = 0
  che = 1
  chf = 0
-> State: 1.2 <-
  e = Code-Available
-> Input: 1.3 <-
  cha = 1
  chf = 1
-> State: 1.3 <-
  f = Fix-Available
  a = Low
```

APPENDIX B

Providing XML Input to CRISNER

Here we provide listings of the SMV models generated by the preference reasoning tool CRISNER (see Chapter 7) for the preference specifications P^{CP}, P^{TCP}, and P^{CPT} introduced in Chapter 1 first, and used in Chapters 2 and 4. The reader will find this a useful reference to build preference specifications that can be used to reason with CRISNER.

1 XML INPUT LISTING FOR P^{CP}

```xml
<?xml version="1.0" encoding="us-ascii"?>
<PREFERENCE-SPECIFICATION>
  <PREFERENCE-VARIABLE>
    <VARIABLE-NAME>a</VARIABLE-NAME>
    <DOMAIN-VALUE>Low</DOMAIN-VALUE>
    <DOMAIN-VALUE>High</DOMAIN-VALUE>
  </PREFERENCE-VARIABLE>
  <PREFERENCE-VARIABLE>
    <VARIABLE-NAME>e</VARIABLE-NAME>
    <DOMAIN-VALUE>Code-Available</DOMAIN-VALUE>
    <DOMAIN-VALUE>Code-Unavailable</DOMAIN-VALUE>
  </PREFERENCE-VARIABLE>
  <PREFERENCE-VARIABLE>
    <VARIABLE-NAME>f</VARIABLE-NAME>
    <DOMAIN-VALUE>Fix-Unavailable</DOMAIN-VALUE>
    <DOMAIN-VALUE>Fix-Available</DOMAIN-VALUE>
  </PREFERENCE-VARIABLE>
  <PREFERENCE-STATEMENT>
    <STATEMENT-ID>p1</STATEMENT-ID>
        <PREFERENCE-VARIABLE>e</PREFERENCE-VARIABLE>
    <PREFERENCE>Code-Available:Code-Unavailable</PREFERENCE>
  </PREFERENCE-STATEMENT>
  <PREFERENCE-STATEMENT>
```

```
    <STATEMENT-ID>p2</STATEMENT-ID>
    <PREFERENCE-VARIABLE>a</PREFERENCE-VARIABLE>
    <PREFERENCE>Low:High</PREFERENCE>
  </PREFERENCE-STATEMENT>
  <PREFERENCE-STATEMENT>
    <STATEMENT-ID>p3</STATEMENT-ID>
    <PREFERENCE-VARIABLE>f</PREFERENCE-VARIABLE>
    <CONDITION>e=Code-Available</CONDITION>
    <PREFERENCE>Fix-Unavailable:Fix-Available</PREFERENCE>
  </PREFERENCE-STATEMENT>
</PREFERENCE-SPECIFICATION>
```

2   XML INPUT LISTING FOR P^{TCP}

```
<?xml version="1.0" encoding="us-ascii"?>
<PREFERENCE-SPECIFICATION>
  <PREFERENCE-VARIABLE>
    <VARIABLE-NAME>a</VARIABLE-NAME>
    <DOMAIN-VALUE>Low</DOMAIN-VALUE>
    <DOMAIN-VALUE>High</DOMAIN-VALUE>
  </PREFERENCE-VARIABLE>
  <PREFERENCE-VARIABLE>
    <VARIABLE-NAME>e</VARIABLE-NAME>
    <DOMAIN-VALUE>Code-Available</DOMAIN-VALUE>
    <DOMAIN-VALUE>Code-Unavailable</DOMAIN-VALUE>
  </PREFERENCE-VARIABLE>
  <PREFERENCE-VARIABLE>
    <VARIABLE-NAME>f</VARIABLE-NAME>
    <DOMAIN-VALUE>Fix-Unavailable</DOMAIN-VALUE>
    <DOMAIN-VALUE>Fix-Available</DOMAIN-VALUE>
  </PREFERENCE-VARIABLE>
  <PREFERENCE-STATEMENT>
    <STATEMENT-ID>p1</STATEMENT-ID>
        <PREFERENCE-VARIABLE>e</PREFERENCE-VARIABLE>
    <PREFERENCE>Code-Available:Code-Unavailable</PREFERENCE>
  </PREFERENCE-STATEMENT>
  <PREFERENCE-STATEMENT>
    <STATEMENT-ID>p'2</STATEMENT-ID>
    <PREFERENCE-VARIABLE>a</PREFERENCE-VARIABLE>
```

```
    <PREFERENCE>Low:High</PREFERENCE>
        <REGARDLESS-OF>e</REGARDLESS-OF>
  </PREFERENCE-STATEMENT>
  <PREFERENCE-STATEMENT>
    <STATEMENT-ID>p3</STATEMENT-ID>
    <PREFERENCE-VARIABLE>f</PREFERENCE-VARIABLE>
    <CONDITION>e=Code-Available</CONDITION>
    <PREFERENCE>Fix-Unavailable:Fix-Available</PREFERENCE>
  </PREFERENCE-STATEMENT>
</PREFERENCE-SPECIFICATION>
```

3   XML INPUT LISTING FOR P^{CPT}

```
<?xml version="1.0" encoding="us-ascii"?>
<PREFERENCE-SPECIFICATION>
  <PREFERENCE-VARIABLE>
    <VARIABLE-NAME>a</VARIABLE-NAME>
    <DOMAIN-VALUE>Low</DOMAIN-VALUE>
    <DOMAIN-VALUE>High</DOMAIN-VALUE>
  </PREFERENCE-VARIABLE>
  <PREFERENCE-VARIABLE>
    <VARIABLE-NAME>e</VARIABLE-NAME>
    <DOMAIN-VALUE>Code-Available</DOMAIN-VALUE>
    <DOMAIN-VALUE>Code-Unavailable</DOMAIN-VALUE>
  </PREFERENCE-VARIABLE>
  <PREFERENCE-VARIABLE>
    <VARIABLE-NAME>f</VARIABLE-NAME>
    <DOMAIN-VALUE>Fix-Unavailable</DOMAIN-VALUE>
    <DOMAIN-VALUE>Fix-Available</DOMAIN-VALUE>
  </PREFERENCE-VARIABLE>
  <PREFERENCE-STATEMENT>
    <STATEMENT-ID>p1</STATEMENT-ID>
        <PREFERENCE-VARIABLE>e</PREFERENCE-VARIABLE>
    <PREFERENCE>Code-Available:Code-Unavailable</PREFERENCE>
  </PREFERENCE-STATEMENT>
  <PREFERENCE-STATEMENT>
    <STATEMENT-ID>p''2</STATEMENT-ID>
    <PREFERENCE-VARIABLE>a</PREFERENCE-VARIABLE>
    <PREFERENCE>Low:High</PREFERENCE>
```

```
        <REGARDLESS-OF>e</REGARDLESS-OF>
        <REGARDLESS-OF>f</REGARDLESS-OF>
  </PREFERENCE-STATEMENT>
  <PREFERENCE-STATEMENT>
    <STATEMENT-ID>p3</STATEMENT-ID>
    <PREFERENCE-VARIABLE>f</PREFERENCE-VARIABLE>
    <CONDITION>e=Code-Available</CONDITION>
    <PREFERENCE>Fix-Unavailable:Fix-Available</PREFERENCE>
  </PREFERENCE-STATEMENT>
</PREFERENCE-SPECIFICATION>
```

APPENDIX C

SMV Models & CTL Queries for Preference Equivalence and Subsumption

Here we provide listings of the SMV models generated by the preference reasoning tool CRISNER (see Chapter 7) for the problem of reasoning about preference equivalence and subsumption (see Chapter 5). The models presented here correspond to the Kripke structures encoding the combined induced preference graphs with respect to preference specifications P^{CP}, P^{TCP}, and P^{CPT} introduced in Chapter 1.

1   SMV MODEL FOR $K(P^{CP}, P^{TCP})$

```
MODULE main

VAR
  f  : {Fix-Available,Fix-Unavailable};
  e  : {Code-Available,Code-Unavailable};
  a  : {High,Low};
  g1 : {0,1};
  g2 : {0,1};
  c__ : {0,1};

FROZENVAR
  a_0 : {High,Low};
  e_0 : {Code-Available,Code-Unavailable};
  f_0 : {Fix-Available,Fix-Unavailable};

IVAR
  cha : {0,1};
  che : {0,1};
  chf : {0,1};

DEFINE
  change := cha=1 | che=1 | chf=1;
  start := a=a_0 & e=e_0 & f=f_0;
```

```
TRANS !change -> c__=1;
TRANS change -> c__=0;
TRANS a_0=next(a_0) & e_0=next(e_0) & f_0=next(f_0);

INIT start=TRUE;

ASSIGN
  next(f) :=
    case
      f=Fix-Available & e=Code-Available & cha=0 & che=0 & chf=1 : Fix-Unavailable;
                      -- #p3 : [e=Code-Available] => f=[Fix-Unavailable:Fix-Available
  ] >> [].
      f=Fix-Unavailable & e=Code-Available & cha=0 & che=0 & chf=1 : Fix-Available;
                      -- #p3 : [e=Code-Available] => f=[Fix-Unavailable:Fix-Available
  ] >> [].
      TRUE : f;
    esac;
  next(e) :=
    case
      e=Code-Unavailable & cha=0 & che=1 & chf=0 : Code-Available;
                      -- #p1 : null => e=[Code-Available:Code-Unavailable] >> [].
      a=Low & cha=1 & che=1 & chf=0 : {Code-Available,Code-Unavailable};
                      -- #p'2 : [] => a=[Low:High] >> [e].
      e=Code-Available & cha=0 & che=1 & chf=0 : Code-Unavailable;
                      -- #p1 : null => e=[Code-Available:Code-Unavailable] >> [].
      TRUE : e;
    esac;
  next(a) :=
    case
      a=High & cha=1 & che=0 & chf=0 : Low;
                      -- #p2 : null => a=[Low:High] >> [].
      a=Low & cha=1 & che=1 & chf=0 : High;
                      -- #p'2 : [] => a=[Low:High] >> [e].
      TRUE : a;
    esac;
  next(g1) :=
    case
      f=Fix-Available & e=Code-Available & cha=0 & che=0 & chf=1 : 1;
      e=Code-Unavailable & cha=0 & che=1 & chf=0 : 1;
      a=High & cha=1 & che=0 & chf=0 : 1;
      TRUE: 0;
    esac;
  next(g2) :=
    case
      f=Fix-Unavailable & e=Code-Available & cha=0 & che=0 & chf=1 : 1;
      a=Low & cha=1 & che=1 & chf=0 : 1;
      e=Code-Available & cha=0 & che=1 & chf=0 : 1;
      a=Low & cha=1 & che=1 & chf=0 : 1;
```

```
      TRUE: 0;
   esac;
```

2   SMV MODEL FOR $K(P^{TCP}, P^{CPT})$

```
MODULE main

VAR
  a : {High,Low};
  e : {Code-Available,Code-Unavailable};
  f : {Fix-Available,Fix-Unavailable};
  g1 : {0,1};
  g2 : {0,1};
  c__ : {0,1};

FROZENVAR
  a_0 : {High,Low};
  e_0 : {Code-Available,Code-Unavailable};
  f_0 : {Fix-Available,Fix-Unavailable};

IVAR
  cha : {0,1};
  che : {0,1};
  chf : {0,1};

DEFINE
  change := cha=1 | che=1 | chf=1;
  start := a=a_0 & e=e_0 & f=f_0;

TRANS !change -> c__=1;
TRANS change -> c__=0;
TRANS a_0=next(a_0) & e_0=next(e_0) & f_0=next(f_0);

INIT start=TRUE;

ASSIGN
  next(a) :=
    case
      a=High & cha=1 & che=1 & chf=0 : Low;
                      -- #p'2 : [] => a=[Low:High]  >> [e].
      a=Low & cha=1 & che=1 & chf=1 : High;
                      -- #p''2 : [] => a=[Low:High]  >> [e, f].
      TRUE : a;
    esac;
  next(e) :=
    case
      a=High & cha=1 & che=1 & chf=0 : {Code-Available,Code-Unavailable};
```

```
                          -- #p'2 : [] => a=[Low:High]  >> [e].
        e=Code-Unavailable & cha=0 & che=1 & chf=0 : Code-Available;
                          -- #p1 : null => e=[Code-Available:Code-Unavailable] >> [].
        a=Low & cha=1 & che=1 & chf=1 : {Code-Available,Code-Unavailable};
                          -- #p''2 : [] => a=[Low:High]  >> [e, f].
        e=Code-Available & cha=0 & che=1 & chf=0 : Code-Unavailable;
                          -- #p1 : null => e=[Code-Available:Code-Unavailable] >> [].
        TRUE : e;
    esac;
  next(f) :=
    case
      f=Fix-Available & e=Code-Available & cha=0 & che=0 & chf=1 : Fix-Unavailable;
                          -- #p3 : [e=Code-Available] => f=[Fix-Unavailable:Fix-Available
  ] >> [].
      a=Low & cha=1 & che=1 & chf=1 : {Fix-Available,Fix-Unavailable};
                          -- #p''2 : [] => a=[Low:High]  >> [e, f].
      f=Fix-Unavailable & e=Code-Available & cha=0 & che=0 & chf=1 : Fix-Available;
                          -- #p3 : [e=Code-Available] => f=[Fix-Unavailable:Fix-Available
  ] >> [].
      TRUE : f;
    esac;
  next(g1) :=
    case
      a=High & cha=1 & che=1 & chf=0 : 1;
      a=High & cha=1 & che=1 & chf=0 : 1;
      e=Code-Unavailable & cha=0 & che=1 & chf=0 : 1;
      f=Fix-Available & e=Code-Available & cha=0 & che=0 & chf=1 : 1;
      TRUE: 0;
    esac;
  next(g2) :=
    case
      a=Low & cha=1 & che=1 & chf=1 : 1;
      f=Fix-Unavailable & e=Code-Available & cha=0 & che=0 & chf=1 : 1;
      a=Low & cha=1 & che=1 & chf=1 : 1;
      a=Low & cha=1 & che=1 & chf=1 : 1;
      e=Code-Available & cha=0 & che=1 & chf=0 : 1;
      TRUE: 0;
    esac;
```

3   SMV MODEL FOR $K(P^{CPT}, P^{CP})$

```
MODULE main

VAR
  f : {Fix-Available,Fix-Unavailable};
  a : {High,Low};
  e : {Code-Available,Code-Unavailable};
```

```
    g1 : {0,1};
    g2 : {0,1};
    c__ : {0,1};

FROZENVAR
    a_0 : {High,Low};
    e_0 : {Code-Available,Code-Unavailable};
    f_0 : {Fix-Available,Fix-Unavailable};

IVAR
    cha : {0,1};
    che : {0,1};
    chf : {0,1};

DEFINE
    change := cha=1 | che=1 | chf=1;
    start := a=a_0 & e=e_0 & f=f_0;

TRANS !change -> c__=1;
TRANS change -> c__=0;
TRANS a_0=next(a_0) & e_0=next(e_0) & f_0=next(f_0);

INIT start=TRUE;

ASSIGN
  next(f) :=
    case
      a=High & cha=1 & che=1 & chf=1 : {Fix-Available,Fix-Unavailable};
                      -- #p''2 : [] => a=[Low:High] >> [e, f].
      f=Fix-Available & e=Code-Available & cha=0 & che=0 & chf=1 : Fix-Unavailable;
                      -- #p3 : [e=Code-Available] => f=[Fix-Unavailable:Fix-Available
    ] >> [].
      f=Fix-Unavailable & e=Code-Available & cha=0 & che=0 & chf=1 : Fix-Available;
                      -- #p3 : [e=Code-Available] => f=[Fix-Unavailable:Fix-Available
    ] >> [].
      TRUE : f;
    esac;
  next(a) :=
    case
      a=High & cha=1 & che=1 & chf=1 : Low;
                        -- #p''2 : [] => a=[Low:High]  >> [e, f].
      a=Low & cha=1 & che=0 & chf=0 : High;
                        -- #p2 : null => a=[Low:High]  >> [].
      TRUE : a;
    esac;
  next(e) :=
    case
      a=High & cha=1 & che=1 & chf=1 : {Code-Available,Code-Unavailable};
```

```
                            -- #p''2 : [] => a=[Low:High]  >> [e, f].
        e=Code-Unavailable & cha=0 & che=1 & chf=0 : Code-Available;
                    -- #p1 : null => e=[Code-Available:Code-Unavailable]  >> [].
        e=Code-Available & cha=0 & che=1 & chf=0 : Code-Unavailable;
                    -- #p1 : null => e=[Code-Available:Code-Unavailable]  >> [].
        TRUE : e;
      esac;
  next(g1) :=
      case
        a=High & cha=1 & che=1 & chf=1 : 1;
        f=Fix-Available & e=Code-Available & cha=0 & che=0 & chf=1 : 1;
        a=High & cha=1 & che=1 & chf=1 : 1;
        a=High & cha=1 & che=1 & chf=1 : 1;
        e=Code-Unavailable & cha=0 & che=1 & chf=0 : 1;
        TRUE: 0;
      esac;
  next(g2) :=
      case
        a=Low & cha=1 & che=0 & chf=0 : 1;
        e=Code-Available & cha=0 & che=1 & chf=0 : 1;
        f=Fix-Unavailable & e=Code-Available & cha=0 & che=0 & chf=1 : 1;
        TRUE: 0;
      esac;
```

4   PREFERENCE SUBSUMPTION QUERY $P^{TCP} \sqsubseteq P^{CP}$ ON $K(P^{TCP}, P^{CP})$

```
-- specification .
      AX ((g1 = 1 & c__ = 1)
            -> EX E [ (g2 = 0 & g1 = 0) U E [ g2 = 1 U (start & g2 = 1) ] ] )
            is false
-- as demonstrated by the following execution sequence
Trace Description: CTL Counterexample
Trace Type: Counterexample
-> State: 1.1 <-'
  a_0 = High
  e_0 = Code-Available
  f_0 = Fix-Unavailable
  e = Code-Available
  f = Fix-Unavailable
  a = High
  g1 = 0
  g2 = 0
  c__ = 0
  start = TRUE
-> Input: 1.2 <-
```

```
  cha = 1
  che = 1
  chf = 0
  change = TRUE
-> State: 1.2 <-
  e = Code-Unavailable
  a = Low
  g1 = 1
  c__ = 1
  start = FALSE
```

5 PREFERENCE SUBSUMPTION QUERY $P^{CP} \sqsubseteq P^{TCP}$ ON $K(P^{CP}, P^{TCP})$

```
-- specification .
      AX ((g1 = 1 & c__ = 1)
             -> EX E [ (g2 = 0 & g1 = 0) U E [ g2 = 1 U (start & g2 = 1) ] ] )
             is true
```

6 PREFERENCE SUBSUMPTION QUERY $P^{TCP} \sqsubseteq P^{CPT}$ ON $K(P^{TCP}, P^{CPT})$

```
-- specification .
      AX ((g1 = 1 & c__ = 1)
             -> EX E [ (g2 = 0 & g1 = 0) U E [ g2 = 1 U (start & g2 = 1) ] ] )
             is true
```

7 PREFERENCE SUBSUMPTION QUERY $P^{CPT} \sqsubseteq P^{TCP}$ ON $K(P^{CPT}, P^{TCP})$

```
-- specification .
      AX ((g1 = 1 & c__ = 1)
             -> EX E [ (g2 = 0 & g1 = 0) U E [ g2 = 1 U (start & g2 = 1) ] ] )
             is false
-- as demonstrated by the following execution sequence
Trace Description: CTL Counterexample
Trace Type: Counterexample
-> State: 1.1 <-
  a_0 = High
  e_0 = Code-Unavailable
  f_0 = Fix-Unavailable
  f = Fix-Unavailable
```

```
      e = Code-Unavailable
      a = High
      g1 = 0
      g2 = 0
      c__ = 0
      start = TRUE
   -> Input: 1.2 <-
      cha = 1
      che = 1
      chf = 1
      change = TRUE
   -> State: 1.2 <-
      f = Fix-Available
      e = Code-Available
      a = Low
      g1 = 1
      c__ = 1
      start = FALSE
```

8   PREFERENCE SUBSUMPTION QUERY $P^{CP} \sqsubseteq P^{CPT}$ ON $K(P^{CPT}, P^{TCP})$

```
-- specification .
      AX ((g1 = 1 & c__ = 1)
              -> EX E [ (g2 = 0 & g1 = 0) U E [ g2 = 1 U (start & g2 = 1) ] ] )
            is true
```

9   PREFERENCE SUBSUMPTION QUERY $P^{CPT} \sqsubseteq P^{TCP}$ ON $K(P^{CPT}, P^{CP})$

```
-- specification .
      AX ((g1 = 1 & c__ = 1)
              -> EX E [ (g2 = 0 & g1 = 0) U E [ g2 = 1 U (start & g2 = 1) ] ] )
            is false
-- as demonstrated by the following execution sequence
Trace Description: CTL Counterexample
Trace Type: Counterexample
-> State: 1.1 <-
   a_0 = High
   e_0 = Code-Unavailable
   f_0 = Fix-Unavailable
   f = Fix-Unavailable
   a = High
   e = Code-Unavailable
```

```
  g1 = 0
  g2 = 0
  c__ = 0
  start = TRUE
-> Input: 1.2 <-
  cha = 1
  che = 1
  chf = 1
  change = TRUE
-> State: 1.2 <-
  f = Fix-Available
  a = Low
  e = Code-Available
  g1 = 1
  c__ = 1
  start = FALSE
```

Bibliography

[1] P. Abdulla, A. Bouajjani, and B. Jonsson. On-the-fly analysis of systems with unbounded, lossy fifo channels. In A. Hu and M. Vardi, editors, *Proceedings of the International Conference on Computer Aided Verification*, volume 1427 of *Lecture Notes in Computer Science*, pages 305–318. Springer Berlin Heidelberg, 1998. 34

[2] S. B. Akers. Binary decision diagrams. *IEEE Transactions on Computers*, 27(6):509–516, June 1978. DOI: 10.1109/TC.1978.1675141. 40

[3] M. F. Ashby. Chapter 8 - case studies: Multiple constraints and conflicting objectives. In M. F. Ashby, editor, *Materials Selection in Mechanical Design (Fourth Edition)*, pages 217 – 242. Butterworth-Heinemann, Oxford, fourth edition, 2011. 47, 52

[4] J. Babar and A. Miner. Meddly: Multi-terminal and edge-valued decision diagram library. In *Seventh International Conference on the Quantitative Evaluation of Systems*, pages 195–196. IEEE, 2010. DOI: 10.1109/QEST.2010.34. 103

[5] C. Baier and J.-P. Katoen. *Principles of Model Checking*. MIT Press, 2008. 38, 40, 43

[6] T. Ball, V. Levin, and S. K. Rajamani. A decade of software model checking with SLAM. *Communications of the ACM*, 54(7):68–76, 2011. DOI: 10.1145/1965724.1965743. 34

[7] K. Bertet, J. Gustedt, and M. Morvan. Weak-order extensions of an order. *Theoretical Compuer Science*, 304:249–268, 2003. DOI: 10.1016/S0304-3975(03)00132-4. 89

[8] D. Beyer and S. Löwe. Explicit-state software model checking based on cegar and interpolation. In *Fundamental Approaches to Software Engineering*, volume 7793 of *Lecture Notes in Computer Science*, pages 146–162. Springer Berlin Heidelberg, 2013. DOI: 10.1109/TC.1978.1675141. 34

[9] A. Biere, A. Cimatti, E. M. Clarke, and Y. Zhu. Symbolic model checking without bdds. In *Proceeding of the International Conference on Tools and Algorithms for the Construction and Analysis of Systems, LNCS 1579*, pages 193–207. Springer, 1999. DOI: 10.1007/3-540-49059-0_14. 64

[10] B. Bollig and I. Wegener. Improving the variable ordering of obdds is np-complete. *IEEE Transactions on Computers*, 45(9):993–1002, Sep 1996. DOI: 10.1109/12.537122. 42

[11] R. Booth, Y. Chevaleyre, J. Lang, J. Mengin, and C. Sombattheera. Learning conditionally lexicographic preference relations. In *Proceedings of the European Conference on Artificial Intelligence*, pages 269–274. IOS Press, 2010. 15

[12] D. Bosnacki and G. J. Holzmann. Improving spin's partial-order reduction for breadth-first search. In *Model Checking Software, 12th International SPIN Workshop*, pages 91–105, 2005. DOI: 10.1007/11537328_10. 34

[13] C. Boutilier, F. Bacchus, and R. I. Brafman. UCP-networks: A directed graphical representation of conditional utilities. In *Proceedings of International Conference in Uncertainty in Artificial Intelligence*, pages 56–64, 2001. 14

[14] C. Boutilier, R. I. Brafman, C. Domshlak, H. H. Hoos, and D. Poole. CP-nets: A tool for representing and reasoning with conditional ceteris paribus preference statements. *Journal of Artificial Intelligence Research*, 21:135–191, 2004. DOI: 10.1613/jair.1234. 74

[15] C. Boutilier, R. I. Brafman, H. H. Hoos, and D. Poole. Reasoning with conditional ceteris paribus preference statements. In *Proceedings of the International Conference on Uncertainty in Artificial Intelligence*, pages 71–80, 1999. 16, 22

[16] C. Boutilier, R. I. Brafman, H. H. Hoos, and D. Poole. Cp-nets: A tool for representing and reasoning with conditional ceteris paribus preference statements. *Joural of Artificial Intelligence Research*, 21:135–191, 2004. DOI: 10.1613/jair.1234. 7, 14, 17, 23, 26, 29, 32, 66

[17] S. Bouveret, U. Endriss, and J. Lang. Conditional importance networks: A graphical language for representing ordinal, monotonic preferences over sets of goods. In *Proceedings of the International Joint Conference on Artificial Intelligence*, pages 67–72, 2009. 7, 14, 20, 22, 25, 32, 66

[18] R. I. Brafman, C. Domshlak, and S. E. Shimony. On graphical modeling of preference and importance. *Journal of Artificial Intelligence Research*, 25:389–424, 2006. DOI: 10.1613/jair.1895. 7, 14, 18, 22, 24, 26, 29, 66

[19] R. I. Brafman, E. Pilotto, F. Rossi, D. Salvagnin, K. B. Venable, and T. Walsh. The next best solution. In *Proceedings of the National Conference on Artificial Intelligence*, pages 1537–1540. AAAI Press, 2011. 77

[20] R. I. Brafman, F. Rossi, D. Salvagnin, K. B. Venable, and T. Walsh. Finding the next solution in constraint- and preference-based knowledge representation formalisms. In *Proceedings of the International Conferences on Principles of Knowledge Representation and Reasoning*, pages 425–433. AAAI Press, 2010. 77

[21] R. E. Bryant. Graph-based algorithms for boolean function manipulation. *IEEE Transactions on Computers*, 35(8):677–691, Aug. 1986. DOI: 10.1109/TC.1986.1676819. 40, 53

[22] J. Burch, E. Clarke, K. McMillan, D. Dill, and L. Hwang. Symbolic model checking: 10^{20} states and beyond. In *Logic in Computer Science*, pages 428–439, Jun 1990. 34, 41

[23] J. Chomicki. Preference formulas in relational queries. *ACM Transactions on Database Systems*, 28(4):427–466, 2003. DOI: 10.1145/958942.958946. 77

[24] J. Chomicki, P. Godfrey, J. Gryz, and D. Liang. Skyline with presorting: Theory and optimizations. In *Intelligent Information Processing and Web Mining*, pages 595–604. Springer, 2005. DOI: 10.1007/3-540-32392-9_72. 77

[25] G. Ciardo, G. Lüttgen, and R. Siminiceanu. Saturation: an efficient iteration strategy for symbolic state space generation. In *Proceedings of the International Conference on Tools and Algorithms for the Construction and Analysis of Systems*, pages 328–342. Springer-Verlag, 2001. DOI: 10.1007/3-540-45319-9_23. 64

[26] A. Cimatti, E. Clarke, E. Giunchiglia, F. Giunchiglia, M. Pistore, M. Roveri, R. Sebastiani, and A. Tacchella. NuSMV Version 2: An OpenSource Tool for Symbolic Model Checking. In *Proceedings of the International Conference on Computer-Aided Verification*, Copenhagen, Denmark, July 2002. Springer. DOI: 10.1007/3-540-45657-0_29. 53, 78

[27] E. Clarke, O. Grumberg, and D. Peled. *Model Checking*. MIT Press, January 2000. 7, 74

[28] E. M. Clarke and E. A. Emerson. Design and synthesis of synchronization skeletons using branching-time temporal logic. In *Logic of Programs, Workshop*, pages 52–71. Springer-Verlag, 1982. DOI: 10.1007/BFb0025774. 34, 35

[29] E. M. Clarke, E. A. Emerson, and A. P. Sistla. Automatic verification of finite-state concurrent systems using temporal logic specifications. *ACM Transactions on Programming Languages and Systems*, 8(2):244–263, 1986. DOI: 10.1145/5397.5399. 34

[30] E. M. Clarke, O. Grumberg, S. Jha, Y. Lu, and H. Veith. Counterexample-guided abstraction refinement for symbolic model checking. *Journal of ACM*, 50(5):752–794, 2003. DOI: 10.1145/876638.876643. 34

[31] E. M. Clarke, O. Grumberg, and D. Peled. *Model Checking*. MIT Press, 1999. 34

[32] V. Conitzer. Making decisions based on the preferences of multiple agents. *Communications of the ACM*, 53(3):84–94, 2010. DOI: 10.1145/1666420.1666442. 105

[33] B. Cook, D. Kroening, and N. Sharygina. Symbolic model checking for asynchronous boolean programs. *Model Checking Software*, 10:75. DOI: 10.1007/11537328_9. 64

[34] C. Domshlak and R. I. Brafman. CP-nets - reasoning and consistency testing. In *Proceedings of the International Conference on Principles of Knowledge Representation and Reasoning*, pages 121–132, 2002. 32

[35] C. Domshlak, E. Hüllermeier, S. Kaci, and H. Prade. Preferences in ai: An overview. *Artificial Intelligence*, 175(7-8):1037–1052, 2011. DOI: 10.1016/j.artint.2011.03.004. 2, 11

[36] J. Doyle and R. H. Thomason. Background to qualitative decision theory. *AI magazine*, 20:55–68, 1999. 1, 2

[37] P. C. Fishburn. Interval graphs and interval orders. *Discrete Mathematics*, 55(2):135–149, 1985. DOI: 10.1016/0012-365X(85)90042-1. 49, 50

[38] S. French. *Decision Theory: An Introduction to the Mathematics of Rationality*. Ellis Horwood Limited, 1986. 1, 105

[39] S. French. Decision theory: An introduction to the mathematics of rationality. 1986. 49

[40] J. Goldsmith and U. Junker. Preference handling for artificial intelligence. *AI Magazine*, 29(4):9–12, 2008. 2

[41] J. Goldsmith, J. Lang, M. Truszczynski, and N. Wilson. The computational complexity of dominance and consistency in CP-nets. *Journal of Artificial Intelligence Research*, 33:403–432, 2008. DOI: 10.1613/jair.2627. 7, 14, 17, 32

[42] O. Grumberg, S. Livne, and S. Markovitch. Learning to order bdd variables in verification. *Journal of Artificial Intelligence Research*, 18:2003, 2003. DOI: 10.1613/jair.1096. 42

[43] D. Gusfield and R. W. Irving. *The Stable marriage problem - structure and algorithms*. Foundations of computing series. MIT Press, 1989. 65

[44] A. Hadjali, S. Kaci, and H. Prade. Database preferences queries–a possibilistic logic approach with symbolic priorities. In *Foundations of Information and Knowledge Systems*, pages 291–310. Springer, 2008. DOI: 10.1007/978-3-540-77684-0_20. 77

[45] S. O. Hansson. What is ceteris paribus preference? *Journal of Philosophical Logic*, 25(3):307–332, 1996. DOI: 10.1007/BF00248152. 16, 22

[46] J. W. Hatfield, N. Immorlica, and S. D. Kominers. Testing substitutability. *Games and Economic Behavior*, 75(2):639–645, 2012. DOI: 10.1016/j.geb.2011.11.007. 65

[47] T. A. Henzinger, R. Jhala, R. Majumdar, and G. Sutre. Lazy abstraction. In *Proceedings of the ACM SIGPLAN-SIGACT Symposium on Principles of Programming Languages*, pages 58–70, New York, NY, USA, 2002. ACM. DOI: 10.1145/565816.503279. 34

[48] G. J. Holzmann. Software model checking with SPIN. *Advances in Computers*, 65:78–109, 2005.

[49] G. J. Holzmann. Mars code. *Communications of the ACM*, 57(2):64–73, Feb. 2014. DOI: 10.1145/2560217.2560218. 34

[50] M. Huth and M. Ryan. *Logic in Computer Science: modelling and reasoning about systems (second edition)*. Cambridge University Press, 2004. 38, 40, 43

[51] A. Jorge, S. A. McIlraith, et al. Planning with preferences. *AI Magazine*, 29(4):25, 2009. DOI: 10.1609/aimag.v29i4.2204. 105

[52] S. Kaci. *Working with Preferences: Less Is More.* Springer-Verlag Berlin Heidelberg, 1st edition, 2011. DOI: 10.1007/978-3-642-17280-9. 15

[53] V. Kahlon, C. Wang, and A. Gupta. Monotonic partial order reduction: An optimal symbolic partial order reduction technique. In *Proceedings of the International Conference on Computer Aided Verification*, pages 398–413. Springer-Verlag, 2009. DOI: 10.1007/978-3-642-02658-4_31. 64

[54] R. L. Keeney and H. Raiffa. *Decisions with Multiple Objectives: Preferences and Value Trade-Offs.* Cambridge University Press, 1993. 1, 105

[55] W. Kießling. Foundations of preferences in database systems. In *Proceedings of the International conference on Very Large Data Bases*, pages 311–322. VLDB Endowment, 2002. DOI: 10.1016/B978-155860869-6/50035-4. 77

[56] D. Kozen. Results on the propositional μ-calculus. *Theoretical Computer Science*, 27(3):333 – 354, 1983. DOI: 10.1016/0304-3975(82)90125-6. 34, 52

[57] T. Lappas, K. Liu, and E. Terzi. Finding a team of experts in social networks. In *Proceedings of the ACM SIGKDD international conference on Knowledge discovery and data mining*, pages 467–476. ACM, 2009. DOI: 10.1145/1557019.1557074. 105

[58] X. Liu and M. Truszczynski. Preference trees: A language for representing and reasoning about qualitative preferences. In *Multidisciplinary Workshop on Advances in Preference Handling*, pages 55–60, 2014. 15

[59] X. Liu and M. Truszczynski. Learning partial lexicographic preference trees over combinatorial domains. In *Twenty-Ninth AAAI Conference on Artificial Intelligence*, pages 1539–1545. AAAI Press, 2015. 15

[60] K. L. McMillan. *Symbolic Model Checking: An Approach to the State Explosion Problem.* PhD thesis, Carnegie Mellon University, Pittsburgh, PA, USA, 1992. 34, 41

[61] K. L. McMillan. Cadence SMV (software). Release 10-11-02p1. Available at: http: //www.kenmcmil.com/smv.html, 2002. 102

[62] O. Morgenstern and J. Von Neumann. *Theory of Games and Economic Behavior*. Princeton University Press, 1944. 1, 49, 105

[63] T. D. Noia, T. Lukasiewicz, M. V. Martinez, G. I. Simari, and O. Tifrea-Marciuska. Computing k-rank answers with ontological cp-nets. In T. Lukasiewicz, R. Peñaloza, and A.-Y. Turhan, editors, *Proceedings of the Workshop on Logics for Reasoning about Preferences, Uncertainty, and Vagueness*, volume 1205 of *CEUR Workshop Proceedings*, pages 74–87. CEUR-WS.org, 2014. 77

[64] T. D. Noia, T. Lukasiewicz, M. V. Martinez, G. I. Simari, and O. Tifrea-Marciuska. *Ontological CP-Nets*, volume 8816 of *Lecture Notes in Computer Science*, pages 289–308. Springer, 2014. 77

[65] NuSMV: A new symbolic model checker. nusmv.fbk.edu. 40

[66] N. I. of Standards and Technology. Common vulnerability scoring system version 2. https: //nvd.nist.gov/CVSS-v2-Calculator, 2015. [Online; accessed 15-June-2015]. 2

[67] M. Osborne. *An introduction to game theory*. Oxford University Press, New York, 2004. 105

[68] Z. J. Oster, G. R. Santhanam, and S. Basu. Automating analysis of qualitative preferences in goal-oriented requirements engineering. In *ASE*, pages 448–451, 2011. DOI: 10.1109/ASE.2011.6100096. 5, 20, 105

[69] Z. J. Oster, G. R. Santhanam, S. Basu, and V. Honavar. Model checking of qualitative sensitivity preferences to minimize credential disclosure. In *FACS*, pages 205–223, 2012. DOI: 10.1007/978-3-642-35861-6_13. 20, 102

[70] D. Peled. All from one, one for all: on model checking using representatives. In *Proceedings of International Conference on Computer Aided Verification*, volume 697 of *Lecture Notes in Computer Science*, pages 409–423. Springer Berlin Heidelberg, 1993. DOI: 10.1007/3-540-56922-7_34. 34

[71] E. Pilotto, F. Rossi, K. B. Venable, and T. Walsh. Compact preference representation in stable marriage problems. In *Proceedings of the First International Conference on Algorithmic Decision Theory*, pages 390–401, 2009. DOI: 10.1007/978-3-642-04428-1_34. 77

[72] M. S. Pini, F. Rossi, K. B. Venable, and T. Walsh. Aggregating partially ordered preferences. *Journal of Logic and Computation*, 19(3):475–502, 2009. DOI: 10.1093/logcom/exn012. 105

[73] A. Pnueli. The temporal semantics of concurrent programs. In G. Kahn, editor, *Semantics of Concurrent Computation*, volume 70 of *Lecture Notes in Computer Science*, pages 1–20. Springer Berlin Heidelberg, 1979. 34

[74] A. Pnueli, J. Xu, and L. Zuck. Liveness with $(0, 1, \infty)$- counter abstraction. In E. Brinksma and K. Larsen, editors, *Proceedings of the International Conference on Computer Aided Verification*, volume 2404 of *Lecture Notes in Computer Science*, pages 107–122. Springer Berlin Heidelberg, 2002. 34

[75] J. Queille and J. Sifakis. Specification and verification of concurrent systems in cesar. In M. Dezani-Ciancaglini and U. Montanari, editors, *International Symposium on Programming*, volume 137 of *Lecture Notes in Computer Science*, pages 337–351. Springer Berlin Heidelberg, 1982. 34

[76] J.-P. Queille and J. Sifakis. Specification and verification of concurrent systems in CESAR. In *International Symposium on Programming*, pages 337 – 351. Springer Verlag, 1982. DOI: 10.1007/3-540-11494-7_22. 7

[77] F. Rossi, K. B. Venable, and T. Walsh. mcp nets: Representing and reasoning with preferences of multiple agents. In *Proceedings of the National Conference on Artificial Intelligence*, pages 729–734. AAAI Press, 2004. 65

[78] F. Rossi, K. B. Venable, and T. Walsh. *A short introduction to preferences : between artificial intelligence and social choice / Francesca Rossi, Kristen Brent Venable, Toby Walsh.* [S.l.] : Morgan & Claypool Publishers, 2011. DOI: 10.2200/S00372ED1V01Y201107AIM014. 105

[79] G. Santhanam and K. Gopalakrishnan. Pavement life-cycle sustainability assessment and interpretation using a novel qualitative decision procedure. *Journal of Computing in Civil Engineering*, 27(5):544–554, 2013. DOI: 10.1061/(ASCE)CP.1943-5487.0000228. 47, 52

[80] G. R. Santhanam. Crisner a qualitative preference reasoner. http://fmg.cs.iastate .edu/project-pages/preference-reasoner/, 2015. [Online; accessed 06-February-2015]. 8, 99

[81] G. R. Santhanam, S. Basu, and V. Honavar. Tcp- compose?–a tcp-net based algorithm for efficient composition of web services using qualitative preferences. In *Proceedings of the International Conference on Service-Oriented Computing*, pages 453–467. Springer, 2008. DOI: 10.1007/978-3-540-89652-4_34. 105

[82] G. R. Santhanam, S. Basu, and V. Honavar. Efficient dominance testing for unconditional preferences. In *Proceedings of the International Conference on the Principles of Knowledge Representation and Reasoning*, pages 590–592. AAAI Press, 2010. 47, 51

[83] G. R. Santhanam, S. Basu, and V. Honavar. Identifying sustainable designs using preferences over sustainability attributes. In *Papers from the Spring Symposium on Artificial Intelligence and Sustainable Design*, 2011. 47, 52, 105

[84] G. R. Santhanam, S. Basu, and V. Honavar. Representing and reasoning with qualitative preferences for compositional systems. *Journal of Artificial Intelligence Research*, 42:211–274, 2011. DOI: 10.1613/jair.3339. 49, 105

[85] The SMV system. `http://www.cs.cmu.edu/~modelcheck/smv.html`. 40

[86] R. Tarjan. Depth-first search and linear graph algorithms. In *Annual Symposium on Switching and Automata Theory*, pages 114–121, Oct 1971. DOI: 10.1109/SWAT.1971.10. 34, 40

[87] A. Tarski. A lattice-theoretical fixpoint theorem and its applications. *Pacific Journal of Mathematics*, 5(2):285–309, 1955. DOI: 10.2140/pjm.1955.5.285. 34

[88] W. Trabelsi, N. Wilson, D. G. Bridge, and F. Ricci. Preference dominance reasoning for conversational recommender systems: a comparison between a comparative preferences and a sum of weights approach. *International Journal on Artificial Intelligence Tools*, 20(4):591–616, 2011. DOI: 10.1142/S021821301100036X. 65

[89] E. Tsang. *Foundations of Constraint Satisfaction: The Classic Text*. BoD–Books on Demand, 2014. 77

[90] M. Y. Vardi and P. Wolper. An automata-theoretic approach to automatic program verification. In *Symposium on Logic in Computer Science*. IEEE Computer Society, 1986. DOI: 10.1007/3-540-50403-6_33. 34

[91] T. Walsh. Representing and reasoning with preferences. *AI Magazine*, 28(4):59, 2007. DOI: 10.1609/aimag.v28i4.2068. 1, 105

[92] F. Wang. Efficient data structure for fully symbolic verification of real-time software systems. In *Proceedings of the International Conference on Tools and Algorithms for the Construction and Analysis of Systems, LNCS 1785*, pages 157–171. Springer-Verlag, 2000. DOI: 10.1007/3-540-46419-0_12. 64

[93] N. Wilson. Consistency and constrained optimisation for conditional preferences. In *Proceedings of the European Conference on Artificial Intelligence*, pages 888–894, 2004. DOI: 10.1016/j.artint.2010.11.018. 22, 51, 66

[94] N. Wilson. Extending CP-nets with stronger conditional preference statements. In *Proceedings of the National Conference on Artificial Intelligence*, pages 735–741, 2004. 7, 14, 22, 24, 29, 51

[95] N. Wilson. An efficient upper approximation for conditional preference. In *Proceedings of the European Conference on Artificial Intelligence*, pages 472–476, 2006.

[96] N. Wilson. Computational techniques for a simple theory of conditional preferences. *Artificial Intelligence*, 175(7-8):1053–1091, 2011. DOI: 10.1016/j.artint.2010.11.018. 14, 19, 26, 32

Authors' Biographies

GANESH RAM SANTHANAM

Ganesh Ram Santhanam is an Associate Scientist at the Department of Electrical and Computer Engineering at Iowa State University. He received his Ph.D. in computer science from Iowa State University in 2010. His research interests include knowledge representation and reasoning, computational decision theory, software engineering, and cyber-security. His doctoral dissertation focused on model checking-based approaches to reasoning with qualitative preferences, and preference reasoning for cyber-defense applications. He has published over 20 research articles on these topics in major journals and conferences in artificial intelligence and software engineering.

SAMIK BASU

Samik Basu is a professor of computer science at Iowa State University. He received his Ph.D. in computer science from the State University of New York at Stony Brook in 2003. His research focuses on formal specification and verification of systems, and the application of logic-based techniques to address safety, security, and optimization problems for software and network-based systems. His research has been funded by several grants from the National Science Foundation. He has published over 70 research articles in major journals and conferences.

VASANT HONAVAR

Vasant Honavar is professor of information sciences and technology and of computer science at the Pennsylvania State University where he holds the Edward Frymoyer Endowed Chair, and heads the artificial intelligence Research Laboratory and the Center for Big Data Analytics and Discovery Informatics. He received his Ph.D. specializing in artificial intelligence from the University of Wisconsin at Madison in 1990. Honavar's current research and teaching interests include artificial intelligence, machine learning, bioinformatics, big data analytics, discovery informatics, social informatics, security informatics, and health informatics. Honavar has led research projects funded by National Science Foundation, the National Institutes of Health, the United States Department of Agriculture, and the Department of Defense that have resulted in foundational research contributions (documented in over 250 peer-reviewed publications) in scalable approaches to building predictive models from large, distributed, semantically disparate data (big data); constructing predictive models from sequence, image, text, multi-relational, graph-structured data; eliciting causal information from multiple sources of observational and experimental data; selective sharing of knowledge across disparate knowledge bases; representing and reasoning about preferences; composing complex services from components; and applications in bioinformatics, social network informatics, health informatics, energy informatics, and security informatics.